SUSTAINABLE ENERGY - RECENT STUDIES

Edited by **Alemayehu Gebremedhin**

Sustainable Energy - Recent Studies
http://dx.doi.org/10.5772/3430
Edited by Alemayehu Gebremedhin

Contributors
Tim Sharpe, Branislav Repic, Dag Henning, Alemayehu Gebremedhin, Denizar Cruz Martins, Abdul Malik, Jakob Rosenqvist, Patrik Thollander, patrik Rohdin, Mats Söderström

Published by InTech
Janeza Trdine 9, 51000 Rijeka, Croatia

Publishing Process Manager Iva Lipovic
Typesetting InTech DTP Team
Cover InTech Design Team

Additional hard copies can be obtained from orders@intechopen.com

Sustainable Energy - Recent Studies, Edited by Alemayehu Gebremedhin
p. cm.
ISBN 978-953-51-0912-9

For EU product safety concerns:
IN TECH d.o.o., Prolaz Marije Krucifikse Kozulić 3, 51000 Rijeka, Croatia,
info@intechopen.com or visit our website at intechopen.com.

Meet the editor

Dr. Alemayehu Gebremedhin is a researcher and senior lecturer at Gjøvik University College (GUC), Norway. He also oversees the study program Renewable Energy at GUC. He is also associate professor in Energy systems at Linköping University, Sweden where he earned his licentiate degree and Ph.D. Dr. Gebremedhin's experience in the area of energy includes system analyses, modelling and optimisation of energy systems, district heating systems, renewable energy and industrial energy conservation. Dr. Gebremedhin is the member of the advisory board of Inland Norway Energy Agency and co-founder of bioenergy alliance inland Norway.

Contents

Preface

Today, access to energy is considered for granted for many, while the situation is the opposite for billions of people who still lack energy for basic needs such as cooking and lighting.

Rising energy demand because of economic growth and also population growth will likely lead to higher energy prices, which in turn can exacerbate the situation for a large part of the world population. Moreover, the environmental impact of energy exploitation and use, which to a large extent is based on fossil energy resources, has already reached a level that calls for greater attention. Therefore, the energy and climate issues have gained increasing attention in the political agenda at different levels both nationally and internationally. This has resulted in various forms of policies that are intended to promote sustainability through increased share of alternative energy sources and efficient energy use.

The fact that the world's energy systems for the most part are built on fossil energy resources and that renewable energy alternative must constantly prove competitive means that it will take time to bring about a drastic change in the global energy map. This is a process where the energy system must be improved step by step towards a sustainable direction. The challenges in the area of resources, energy supply security, economic growth and the environment require a joint action in which the development of alternative energy, technology improvement and introduction of relevant policy instruments should go hand in hand in order to achieve sustainable development at all levels.

This book is a collection of six interesting chapters which focus on some specific areas that are vital for the promotion of sustainability in the area of energy and environment. The papers which are included in this book can be generally classified under the following themes: district heating, bioenergy, solar energy, industrial energy auditing and building. The book will provide a general but also some specific insight into the different topics mentioned above.

The first chapter describes the value of district heating and cooling as a means toward efficient utilisation of energy resources. Here, important aspects such as industrial surplus heat utilisation, poly-generation opportunity, biomass fuels utilisation and energy-carrier switching are discussed.

Bioenergy is the main theme in chapter 2 and 3. Chapter 2 gives an extensive literature review on potential and use of bioenergy in ASEAN countries. Moreover, the importance of clean development mechanism (CDM) to boost bioenergy projects in these regions is highlighted. Chapter 3 deals with combustion technology development where local biomass (agricultural residues) can be used for local heat production. This chapter is supported by experiments where the "cigar" firing technology is investigated for combustion of biomass bales.

Chapter 4 deals with the analysis of photovoltaic Maximum Power Point Trackers. This is vital since the output voltage and current from PV modules depend on environmental condition such as solar radiation and temperature. Here, a converter and a tracking algorithm which enables an optimal operation of photovoltaic systems is proposed.

The remaining two chapters focus on industry and residence building with the aim to achieve increased sustainability. In chapter 5 a methodology for industrial energy audit with measurement examples are presented, while the last chapter describes case studies where Building Performance Evaluation is used (here the focus is indoor climate verse energy use).

The search for energy supply security, economic growth and an energy system with minimum impact on the environment underscores the importance of sustainable energy as never before.

The purpose of this book is not to provide answers to how to achieve a sustainable energy system, but rather an effort that leads in that direction by presenting a series of research. The book presents a series of research results ranging from theoretical studies to advanced technological solutions to improve the use of some of the renewable energy sources.

Alemayehu Gebremedhin
Sustainable energy group
Department of Technology, Economy and Management
Gjøvik University College
Norway

Chapter 1

District Heating and Cooling Enable Efficient Energy Resource Utilisation

Dag Henning and Alemayehu Gebremedhin

Additional information is available at the end of the chapter

http://dx.doi.org/10.5772/51837

1. Introduction

Economic development in transition countries, such as China and India, increase global energy use. Therefore, the demand for energy carriers grows, which should increase energy prices. Global energy supply is dominated by fossil fuels, such as coal, oil and natural gas, and this situation is likely to remain for many years even if the use of renewable energy sources (e.g., biomass, solar energy and wind energy) is expanding. Higher energy prices make certain changes of the energy system more profitable: use of *free* energy sources, such as sun and wind, efficiency improvements of energy supply, as well as energy conservation measures, which reduce energy use.

Several policies on various levels now promote increased utilisation of renewable energy sources and reduced energy end-use, for example in buildings. But there are also comprehensive systems that link energy resources with demand for energy. *District heating* is such a concept, which is common in many countries where space heating of buildings is required, for example Iceland, Latvia and Denmark. In a district heating system, heat is distributed through a network of hot-water pipes from heat-supplying plants to heat consumers in a single block or a whole city. The heat is mostly used for space heating and domestic hot water. District-heating systems range from a single development to city-wide networks. *District cooling* works in the corresponding way. *District energy* includes district heating and district cooling. District heating is sometimes called community heating, especially in the UK.

More than one-fourth of the primary energy supply in Europe becomes losses by energy conversion, mainly as heat that is wasted by electricity generation in condensing power plants. These losses are of the same magnitude as the European heat demand [1]. District heating is a means to utilise such losses, which otherwise are wasted, to cover demand for various kinds of heat and even cooling. District heating helps us utilising large amounts of heat that now are wasted in Europe.

Thus, district heating is not only a technology for energy distribution but it increases the amount of available energy resources. District heating can utilise energy sources that are difficult to use for individual buildings, such as unrefined biomass fuels, heat from waste incineration, heat from electricity generation in combined heat and power (CHP) plants and industrial surplus heat, for example heat from pulp and paper mills or production of automotive biofuel. Little of this energy could be utilised without district heating. Therefore, district-heating expansion may be beneficial for economy and environment.

District heating is used for heat supply to various kinds of buildings in villages and cities, primarily multi-family buildings and service premises, where the heat is used for preparation of domestic hot (tap) water and for space heating, normally, through a central waterborne heating system for the whole building.

District heating systems connect energy sources and energy users. District heating can provide affordable energy to consumers by using low-cost energy sources, such as surplus heat and waste. Many of these heat sources can be of local origin and promote local business and industry. What is the most suitable solution depends on the local conditions. By using various energy sources, district heating becomes a central component for waste management systems, forestry, power production and efficient energy use in industry.

1.1. Heat supply

Heat sources that cannot be used for separate houses can in a district-heating system be complemented by technologies that also are applicable at smaller scale, for example, fossil fuels, solar energy and electric heat pumps upgrading low-temperature heat. A small district-heating system can have one or two heating units, whereas a large system can host many different heat sources where, for example, a CHP plant fed with low-cost waste covers the base load throughout the year, a wood-fired heat-only boiler supplies most of the space-heating demand in winter and a boiler using expensive oil covers the peak load during the coldest days.

Base-load plants typically have a low heat production cost but require large investments. The low operation cost makes them suitable for being used during many hours a year. Benefitting from a lower heat cost than from other units pays back the heavy investment. Common base-load supply comes from CHP plants, waste incineration and industrial surplus heat. Oil-fired boilers, on the other hand, have low capacity costs but high operation costs, which make them suitable for covering short periods of peak heat demand.

Figure 1 shows how heat production can take place in a Swedish district heating system during a year. In summer, heat demand and production are low because there is primarily need for heating of domestic hot tap water only but in winter heat production is much larger due to high space heating demand. The base load is covered by industrial surplus heat throughout the year because it has the lowest cost. The higher load in winter is mainly covered by wood used in CHP plants and boilers but fossil CHP production and heat pumps are also used. Some oil is used in heat-only boilers when it is very cold. The units used at high demand have higher heat production costs and are generally more polluting than the plants used at lower demand. Therefore, the marginal cost for district heating production varies in a similar way as the heat demand during the year.

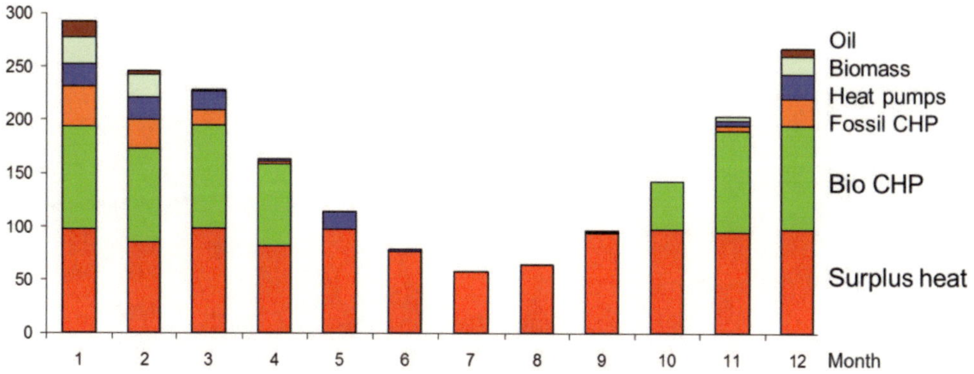

Figure 1. District heating production in a Swedish system (GWh)

The fossil-fuel-fired CHP plant and the heat pumps in the system in Fig. 1 were once built as plants covering the base load but later the wood-fired CHP plant was built, which could produce heat at lower cost and the annual utilisation times for the older plants were decreased. The introduction of industrial surplus heat reduced the use of all other plants to their present levels. District heating demand and production are often shown with a duration curve, which represents heat demand in descending order from the coldest winter days to the warmest summer nights (see e.g., [2]).

1.2. District heating in Sweden

District heating is used extensively in Sweden. Sweden has nine million inhabitants. Fifty annual TWh of district heating cover one-half of the heat market. There is a district heating system in every municipality with more than 10 000 inhabitants and in total there are more than 400 systems. One-half of Swedish district heating is supplied to multi-family houses, the rest mainly to premises, such as schools and offices, and small but growing fractions to industry and single-family houses [3]. House owners chose whether to connect to a district-heating grid or not.

Figure 2 shows the energy sources used for district-heating supply in Sweden since 1970 [4]. The total supply varies between cold and warm years. The last year (2009) was a very cold year. The fuel use for district-heating production in Sweden has switched from almost only using oil in the 1970s to a present mixture with many heat sources. Now, two-thirds of Swedish district heating is produced from wood and waste fuel. Sweden utilises much industrial surplus heat compared to most countries and heat pumps take heat from sewage water and lakes. Minor quantities of biogas and gas from ironworks are used but fossil fuels now produce less than 15% of the district heating (Fig. 2). The fossil carbon-dioxide emissions from district heating have been reduced significantly during the past decades because fossil fuels have produced a decreasing fraction of the heat. This transition has been facilitated by an early introduction of a carbon-dioxide tax and other policy measures [5].

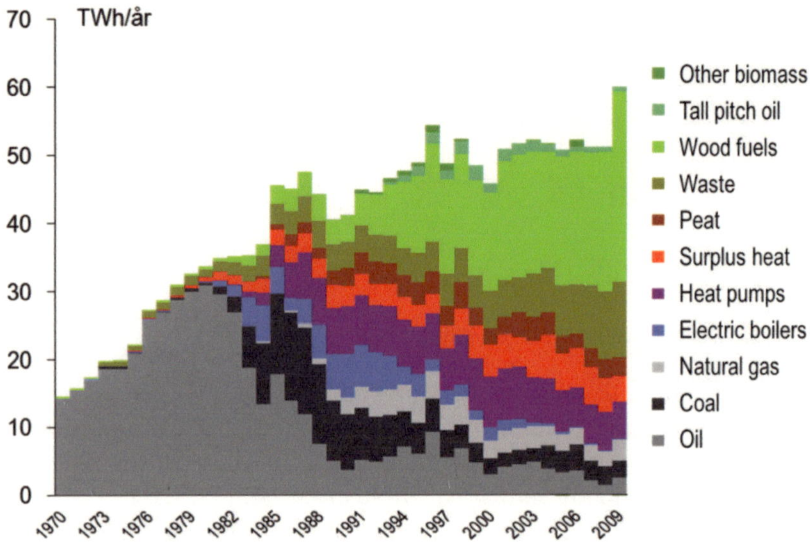

Figure 2. Fuels etc. used for district-heating production in Sweden (TWh/year)

Oil and coal use decreased during the 1980s (Fig. 2) due to increasing taxes. There has been a carbon-dioxide tax in Sweden since 1991, which now is 100 euro/ton. An energy tax was introduced even earlier. There is natural gas only in south-west Sweden. The use of electric boilers for district-heating production increased when nuclear power expanded during the 1980s but decreased when the electricity was taxed in the 1990s. Use of biomass (e.g., woodchips) was first promoted by the taxes on fossil fuels and later also by green electricity certificates and higher electricity prices that make biomass-fired CHP plants more profitable. Waste incineration increases (Fig. 2) because it is prohibited to dispose combustible fuel on a dump and district heating companies collect revenues for taking care of the waste. There have also been investment subsidies to some selected projects using local energy sources. The district-heating increase during the last decades reflects a political commitment to invest in infrastructure and reduce dependency on imported fossil fuels.

http://dx.doi.org/10.5772/51837

2. Methods

Favourable comprehensive solutions can be elucidated through system analysis and optimisation models. These methods can show the best way to use resources to satisfy aims. Common aims are low costs and low environmental impact, which often can be conflicting. The essential features of an issue under study for a system can be represented in a model. Models often help system understanding and reveal relations among components, such as between district-heating production, solar energy extraction and wall insulation. The best solution according to a criterion and under certain conditions can be shown by an optimisation model. For energy issues, an energy system optimisation model can be used to find the best design and operation of a system. In such models, many technical components can be described [6].

Examples of energy system optimisation models are MARKAL (e.g., [7]) and TIMES (e.g., [8]). These models were primarily developed for national energy-system analyses, whereas the model MODEST was originally made for optimising district-heating supply under consideration of heat-demand fluctuations.

MODEST is an energy system optimisation model, which uses the optimisation method linear programming to find the minimum cost for satisfying energy demand and presents the system design and operation that achieves the lowest cost. A large number of options for energy supply and conservation can be considered with this model framework. The user can make a comprehensive representation of the energy system under study with chosen level of detail. Many different energy systems can be analysed as long as the important properties of the system can be described by linear relations. An almost arbitrary set of parameter values may be attributed to each component and energy flow in a system. A flexible time division makes it possible to reflect diurnal, weekly, monthly, seasonal and long-term variations of, for example, costs, capacities and demand. The modelling result presents the optimal investments and the optimal operation of existing and new units as well as emissions and costs [6].

MODEST has been most used for optimisation of electricity and district-heating production. MODEST has been applied to more than 50 district-heating systems, some regional energy systems and a few national power systems. Studied issues include introduction of waste incineration and combined heat and power production and connections between industrial and municipal energy systems [6], for example, how large CHP plant should be built, is waste or biomass the best fuel and should industrial surplus heat be utilised? The model has been used a lot to study impact of energy prices and policy instruments on investments and operations in energy conversion, for example, how emissions allowances influence combined heat and power production.

3. Favourable energy sources and plants

District heating can use heat resources that are more or less impossible to supply to and convert to heat in single houses. Such energy sources include surplus heat from industries, heat produced through combustion of waste and unrefined biomass, as well as heat from larger combined heat and power plants.

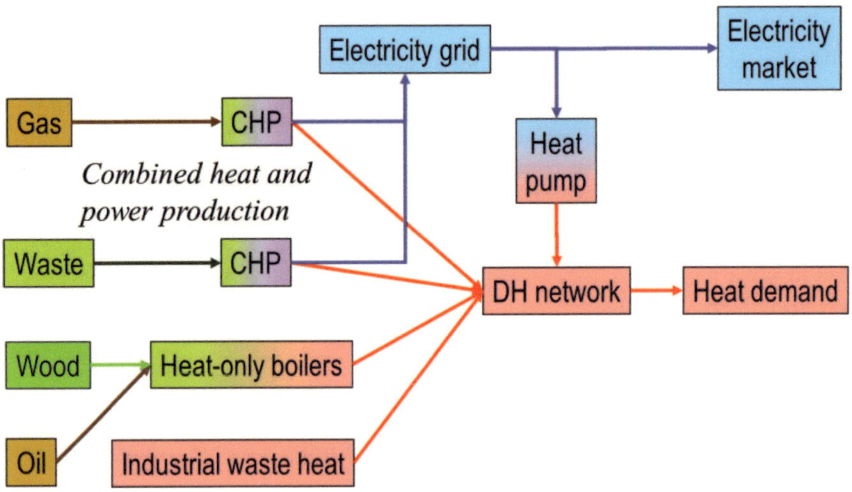

Figure 3. Common plant types and energy flows for district-heating supply at a local Swedish energy company

Figure 3 shows the district heating system in the second largest Swedish city Göteborg (Gothenburg). There is a large city-wide district-heating network with a large heat demand, which means that the system can host many different forms of heat supply. There are two combined heat and power plants, which produce electricity and heat; a natural-gas-fired combined-cycle unit and a waste-fired steam-cycle plant. Industrial surplus heat is bought from two oil refineries. Wood and oil-fired boilers, as well as electric heat pumps produce heat only. The heat is distributed through the district-heating network to the consumers to the right in Fig. 3. This is one example of how district heating enables efficient energy utilisation, but the system also includes components that may be considered less sustainable, such as fossil fuels.

CHP plants offer better fuel use. In condensing power plants, most of the fuel energy is normally wasted. Electricity produced in CHP plants, which produce district heating, can displace electricity from condensing power plants. Due to higher efficiency, less fuel is needed for the electricity generation in CHP plants than condensing power plants because the major fraction of the fuel yields district heating. Therefore, the carbon-dioxide emissions caused by the power production are lower for a CHP plant even if it is fed with fossil fuel. But the environmental benefit is of course even larger if renewable fuels are used for the CHP production.

http://dx.doi.org/10.5772/51837

3.1. Using local renewable energy resources

Available local renewable energy supplies influence the suitability of various solutions at a location. In some places, solar energy may contribute to district-heating supply. But most useful for district heating are biomass fuels and combustible waste.

Biomass fuels can be derived from forestry and agriculture. Use of biomass fuels can initiate local biomass industry and promote local business and development. The demand for biomass fuel from a district-heating plant can make entrepreneurs develop supply systems for, for example, wood fuels, such as tree branches from forests.

Today, much waste is landfilled. Using the waste as fuel or for energy extraction in other ways reduces the landfilling. Waste is a resource and various waste fractions should be separate to make it possible to use them in the most suitable way. The separation can take place at the source, that is, in households and firms, etc. For example, biodegradable waste can yield biogas, which can be used as automotive fuel, whereas other combustible waste (e.g. from households or building demolition) can be used as fuel for electricity and district-heating production.

Utilisation of local renewable energy resources means higher security of energy supply and lower dependency on fuels from other regions and countries. Use of local fuels and generation of electricity reduce the energy bought from other places and increases the money that can create wealth locally. Switching from fossil to renewable fuels also reduces emissions of fossil carbon dioxide and decreases the local contribution to global warming.

However, biomass may be a too valuable resource for producing only heat, which is an energy form of lower value than electricity and automotive fuel. Biomass could rather produce combinations of different energy carriers, such as heat, steam, electricity, automotive fuel and cooling (Sect. 6). Biomass certainly is a renewable energy source but its extraction must anyhow be carried out in a sustainable way. Land also produces food and raw material and a dilemma is that the wealthy can pay more for fuel than the poor can pay for food.

Waste is of renewable, as well as fossil origin. Waste volumes are normally increasing with economic growth but less waste than now should be generated in a sustainable society. Waste incineration may therefore partly be seen as a transition technology rather than being a basis to the present extent in a long-term system.

4. Utilising industrial surplus heat

There is today a huge amount of surplus heat within the energy and industry sectors. In many cases, this surplus heat is not utilised as it should be. This situation is not good since the industrial sector accounts for more than 30% of the world total final energy consumption [9]. This should also be seen in connection to the fact that the total primary energy supply is highly dominated by fossil fuels. Thus, utilising surplus heat is an essential measure to achieve an overall sustainable energy system in a community or region.

Heat can be recovered for repeated utilisation at decreasing temperature levels in industry and finally for space heating. Although it is known that there is surplus heat in certain facilities, in some cases little is known concerning the volume and the quality of the heat. Identifying and measuring surplus heat resources would therefore be necessary before any evaluation can be made. For instance, a study based on energy auditing showed that as much as 500 GWh/year heat energy of different quality is wasted in a single pulp and paper mill [10]. However, knowing heat quantity and quality does not automatically mean that the surplus of heat can be used. Other factors, such as time of availability, heat demand, infrastructure, technology and costs, play decisive roles in determining if the surplus heat can be utilised or not.

District heating offers an outstanding opportunity to utilise surplus heat which otherwise would be wasted. Though the share differs from country to country, the Nordic district heating systems are good examples of using surplus heat [11]. From a consumer's perspective, district heating systems with significant share of surplus heat in its fuel mix offer relatively low heating costs to their consumers. In some municipalities in Sweden, the rather low heat costs can be attributed to surplus heat supply from industries. Availability of surplus heat during summer when the heat demand is low opens an opportunity to produce district-heating-driven cooling for buildings during summertime (Sect. 5).

Utilising surplus heat in district-heating applications is not quite easy since such an endeavour has different issues that need to be resolved. One of the main issues is how to bring about a co-operation platform between players where the use of industrial surplus heat is understood in the light of a broader system perspective. In this case, a municipality, or a region with several municipalities, can be the system boundary when considering energy co-operation. In regions where district heating is well established and where there is relatively high concentration of industrial activities, there might be a need to develop a regional heat market to encourage efficient utilisation of energy resources. Though the core business of an industry is not selling surplus heat, such a market could be a driving force which in turn enables players to take measures that might promote the use of surplus heat. This would mean that industries with substantial amounts of surplus heat can play significant roles as heat suppliers in local or regional markets. This is also important when considering investments in new generation facilities for electricity, steam and heat. In this case, regional energy system optimisation would be valuable to maximise the efficiency of resource utilisation.

Thorough studies that focused on this subject are given in [12,13]. In both studies, several district-heating systems and industrial energy systems of a region are considered and the MODEST model is applied (Sect. 2). The second study [13] was more in detail and it includes several scenarios where measures, such as investment in new facilities, process integration and energy efficiency measures, are considered. In general, both studies indicate an overall system benefit of connecting the various energy systems in forms of reduced total cost, efficient use of plants and reduced carbon-dioxide emissions, but the latter depends strongly on the carbon accounting method applied. Interesting to note is that an enlarged system boundary, encompassing all the district-heating systems and industrial energy systems, enables efficient utilisation of surplus heat that is available within the system. De-

http://dx.doi.org/10.5772/51837

pending on the prevailing conditions, utilisation of surplus heat can be in conflict with the use of CHP. A widened system boundary with a possible heat market may enable the use of both surplus heat and CHP.

Utilising surplus heat should be promoted in similar way as renewable energy where policy instruments are deployed to encourage power production based on renewable sources. Furthermore, a suitable co-operation platform needs to be created where different issues concerning the utilisation of surplus heat in district heating systems can be resolved. However, surplus heat supplies available for district-heating systems may be somewhat reduced by increased heat reuse within industries.

5. Cooling

Cooling of rooms now increases due to higher comfort requirements in, for example, countries in Northern Europe, as well as from the middle class in transitions countries where the living standard is rapidly rising. The desire to cool the rooms is also enhanced by the global warming and the growing number of electric appliances that supply waste heat in the rooms [14].

Normally, electricity-driven refrigeration machines are used to produce cooling. But district-heating sources can also produce cooling for indoor climatisation in absorption cooling machines, either in central plants (e.g., waste incineration plants) supplying a district cooling network or in distributed units situated in the buildings that are to be cooled, which are fed from the district-heating network (Fig. 4). Such solutions mean that electricity is not required for the cooling. It increases the low heat demand during summer and also the electricity generation in CHP plants if the heat is produced there. Thus, cooling can become a basis for electricity generation instead of consuming electricity [2]. Absorption cooling is most suitable when a low-cost fuel, such as waste, can be used.

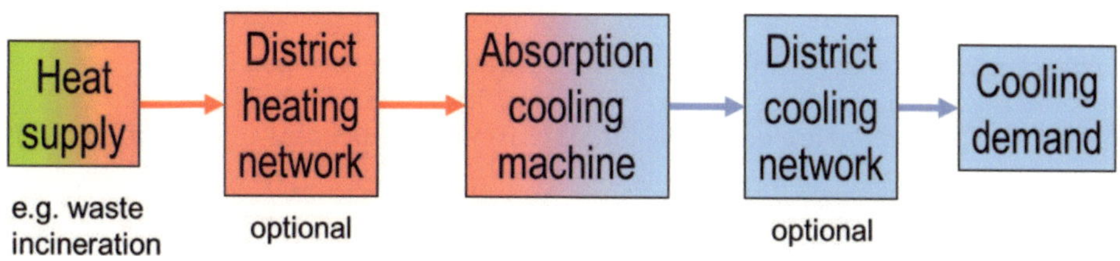

Figure 4. Cooling with heat through absorption cooling. One of the networks is required.

6. Poly-generation of several energy forms

European polices aim at increasing the use of automotive biofuel. Such fuels can be favourably produced in poly-generation plants that can turn various forms of biomass into automotive biofuel, electricity, steam, heat and cooling, which partly is used within the processes but largely is output from the plant. Similar arrangements can be made for other types of industry. A combined heat and power plant can produce district heating, electricity and district cooling as well as steam, which is supplied to an industry. The joint generation of several energy carriers increases the utilisation of installed capacity, increases revenues from delivered energy and may make plant investment more profitable.

The deployment of combined heat and power production in district heating systems connects the heat and power sectors in such a way that the overall production efficiency will be improved substantially. Poly-generation plants that produce automotive fuel connect the stationary energy system with transportation, like electric cars and trains do, and increase the number of options for biomass utilisation and transport provision. The linkage with the transport sector is especially important since the electricity, heat and transport sectors cause more than 60% of globally generated carbon-dioxide emissions from fuel combustion [9]. A combined action within these three sectors will definitely reduce emissions. In this aspect, biomass is a vital resource to meet energy and environmental targets. Using biomass just for heating purposes could be a step toward sustainable development particularly in areas where non-renewable sources are used now. However, other technologies, such as co-generation (CHP), tri-generation (CHP + cooling), and poly-generation, should be considered to maximise the benefit of using biomass. This is especially important in areas where the deployment of CHP is difficult due to barriers, such as insufficient heat load, unfavourable power prices, high investment costs, lack of infrastructure etc. Furthermore, low heat-demand periods are a challenge in district-heating systems with CHP as base load production. This situation can possibly worsen with increasing efficiency within the residence sector where lower heat demand is expected (Sect. 8.1). There is also a desire to cut heat production costs through additional revenues from sales of electricity and automotive fuel since district heating is not always the cheapest alternative in some places. With this background, the poly-generation concept can be helpful for tackling the mentioned issues.

Studies indicate that there are economic and environmental benefits of applying poly-generation concepts. For instance, increased power production from CHP plants can be achieved by integrating lignocellulosic ethanol plants with district heating [15]. Another similar study uses the MODEST model (Sect. 2) to show that a poly-generation configuration would result in lower production cost for heat and reduced emissions as a result of integration [16]. There are also other poly-generation applications where products, such as steam, electricity, heat and wood pellets are generated simultaneously. Such plants are already available, for instance, in Sweden and Norway. Revenues obtained from sales of, primarily, power and vehicle fuel together with renewable incentives seem to encourage the use of biomass resources efficiently and thereby create a favourable condition for the competitiveness of district heating and biomass-based power production.

7. A valuable infrastructure

District heating plants and networks have low operation costs when using low-grade energy resources but they require large initial investments. The cash flow is negative for some years during the establishment of a new district-heating system and the payback time can be rather long, which makes financing more difficult. A long-term perspective on profitability and business models with low risks are essential for the deployment and modernisation of district heating systems. Prevailing public policy support may also be needed to facilitate the development of district heating infrastructure, like for other large-scale systems. Due to the heavy investments made, existing district heating systems are valuable assets, but some systems may require substantial improvements [17].

The district-heating value chain goes from fuel through heat production and distribution to consumer. Most Swedish district heating companies encompass all central parts of this chain, that is, heat production, distribution and sales, which enables utilisation of operational synergies. This arrangement can be favourable because if many actors are involved, a series of agreements are required, which increase business risks, which in turn makes financing more expensive, which may make investments unprofitable [17].

Heavy investments, such as waste-incineration and CHP plants, require a certain size to be profitable and therefore they also need a large district heating system to be suitable. Such a district-heating network may sometimes be achieved through connection of smaller systems.

8. Heat demand

District heating demand may be seen as a valuable resource itself because it enables the utilisation of energy resources that without this demand would be difficult to use. The district-heating demand also makes combined heat and power production possible.

District heating is more suitable the larger the heat load density is (i.e., heat demand per ground area, e.g. [18]) because more heat can be delivered per meter of pipe buried in the ground and network costs can be spread on a larger energy amount. Therefore, district heating is primarily used in larger buildings, for example multi-family residences and service premises, such as hospitals, schools and larger office buildings. But the heat load density that is required for district heating to be economically favourable depends on the heat production cost [2]. If the *heat sink* that district-heating users constitute enables power production or waste reception that yield revenues, it is profitable to build district-heating grids in areas with lower heat load density than if biomass or oil is used to produce the district heat separately.

In some places, heat prices vary in a similar way as the heat production cost during the year (Sect. 1.1) to give consumers a signal on when it is most desirable that they reduce their heat demand. Houses with district heating may, for example, be less suitable for solar heating because district heating often comes from surplus resources, such as waste or waste heat, when there is most solar radiation.

8.1. Lower demand

Now, heat demand is decreasing due to higher outdoor temperatures caused by the enhanced greenhouse effect, as well as policies that promote low-energy houses, which makes district heating a less suitable form of heat supply. All new buildings in the European Union are supposed to be *nearly-zero-energy* buildings in 2020 [19]. Low-energy houses often have thick wall and attic insulation, windows transmitting little heat, ventilation with heat recovery and solar heating. These more advanced installations cause higher investment costs but the lower energy use reduces operation costs.

Lower heat demand should reduce the use of natural resources, such as fossil fuels, and enable biomass to be used for other purposes than space heating, such as production of automotive fuel. But the heat demand reductions are a challenge for district heating and therefore also for the possibilities to utilise energy sources that need district heating to be used, such as industrial surplus heat. Therefore it is important to analyse the interplay between energy supply and energy conservation and between district-heating companies and buildings.

Energy-efficiency measures, such as improved wall insulation and better windows, primarily reduce heat demand in winter and, hence, decrease seasonal demand variations. This may be favourable from a heat-production viewpoint because high-load plants are needed less but base-load plants (Sect 1.1) may be used more, which would reduce operation costs and environmental impact. But base-load plants would also be affected, which could decrease efficient electricity generation in combined heat and power plants.

Åberg and Henning [20] studied the impact of a potential heat-demand reduction due to extensive energy-efficiency measures in existing buildings on district-heating and electricity production by using the energy system optimisation model MODEST (Sect. 2). In the Swedish city under study, the heat-demand reductions would primarily decrease heat-only production, whereas CHP production would be less reduced. The *electricity-to-heat output ratio* for the system would even increase, that is, generated electricity per unit of delivered district heating would increase. Local carbon-dioxide emissions would be lowered by the energy-efficiency measures because less fossil fuel would be used. Global carbon-dioxide emissions would also be reduced though less efficient coal-fired condensing power plants would need to replace the electricity that can no longer be produced in the CHP plants due to reduced heat sink in the buildings. However, only the existing electricity and district-heating production plants are considered in this study [20], whereas a process of gradual heat-demand reduction in present houses would run in parallel with a restructuring of the heat supply system probably including a transition to even larger use of renewable fuels. In such a future system, energy-efficiency measures might not reduce carbon-dioxide emissions.

In a similar study of another city [21], the combined effect of energy-efficiency improvements in existing multi-family buildings and the connection of new low-energy multi-family houses to the district-heating grid was studied with MODEST. These changes would not affect global carbon-dioxide emissions if there is interplay with coal-fired condensing power plants. But heat production plants and fuels used have crucial importance for the environmental impact of district heating. In this case, the heat demand changes would, for example, decrease the use of a CHP plant fuelled with carbon-rich peat, which cause similar carbon-dioxide emissions as coal. The larger impact on CHP production compared to the previously mentioned study is also shown by an electricity-to-heat output ratio for the system that declines with heat demand [21].

8.2. Using district heating at low demand

To make it favourable to use district heating in areas with low heat demand and, thus, to enable utilisation of the energy sources that can only be used through a district-heating system, as much district heating as possible should be used in such an area while still using the heat efficiently.

Besides the traditional purposes domestic hot water and space heating, district heating can be used for industrial processes and all heat supply to household appliances (e.g. dish washers, washing machines, tumble dryers and towel dryers), which now often, at least partly, use electricity for heating. Solar rooftop energy extraction could yield electricity instead of heat, because the latter would reduce the demand for district heating supply.

Henning [22] outlined scenarios for more sustainable energy supply for a development in a Swedish town. In two cases, the buildings were supplied by district heating. In one of these scenarios, district heating was used only in the traditional way for domestic hot water and space heating in normal, but not inefficient, houses. In the other scenario, there were low-energy buildings where district heating also was used for household appliances. Energy that in the first case only disappeared out of the buildings was in the other case utilised for heat supply to dish washers, washing machines, tumble dryers and towel dryers. The first scenario meant more climate-dependent space-heating demand partly covered by expensive high-load fuels (mainly forest wood chips in this town), whereas the latter scenario included more base load in the household appliances, which could be covered by fuels with lower costs (wood waste, [22]).

Many industrial processes have heat demand that partly or wholly can be covered by district heating but now is supplied through fuel or electricity. When required, district heating can be supplemented by boilers to obtain desired high temperatures. Heat demand in industrial manufacturing processes is often more or less independent on outdoor temperature and only has little seasonal variations (besides holidays) in the same manner as domestic hot water. Industrial processes can, therefore, constitute a base demand, which favourably could be covered by base-load plants, such as waste incineration or CHP plants [23].

With lower heat demand, the temperature in district-heating networks can be lower, which means that surplus heat of lower temperature can be utilised. Also, more electricity can be produced in CHP plants because the heat that is extracted after electricity generation can be of lower temperature. A larger fraction of fuel energy can yield high-quality electricity instead of low-quality heat.

9. Useful electricity use

Electricity is widely used for purposes there district heating or cooling could be used instead. Electric heating is used extensively in Norway (Fig. 5) but also in several other countries. Electricity is also generally used for air conditioning. Switching from electric heating or cooling to district heating or cooling naturally reduces the electricity consumption but if the district energy comes from a CHP plant the switching may also enable a larger electricity production there and less other electricity production is needed, which often is coal-fired condensing power plants. Thus, such energy-carrier switching influences the power system twofold through reduced demand and changed generation, which can lower fuel consumption and carbon-dioxide emissions. In Sweden, the use of district heating could be increased by 25% if all electric heating in non-rural areas was replaced by district heating [14].

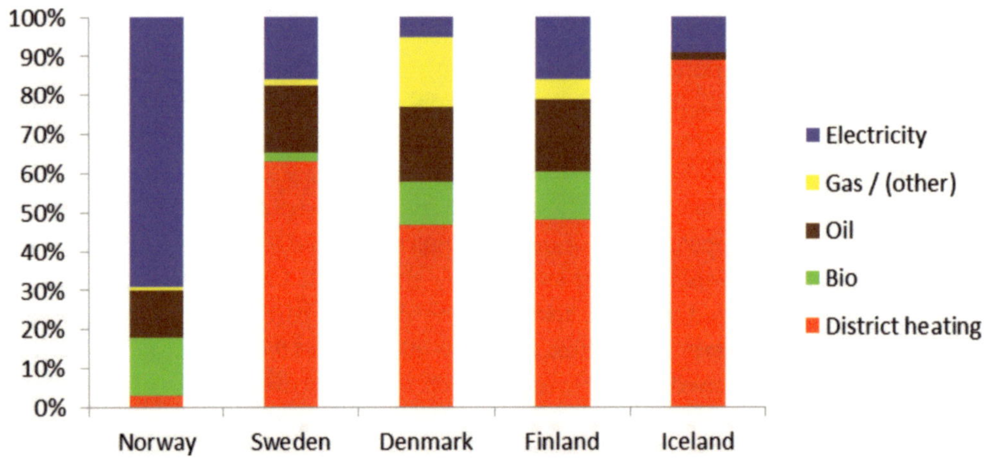

Figure 5. Nordic heat market (Source: Norsk Bioenergiforening (NoBio)

Seen from an exergy point of view, electricity should not be used for heating purposes. This is particularly obvious if electricity is generated with low efficiency and with fuels that are both costly and not environmentally friendly. On the other hand, it could be more difficult to argue against using electricity if the power is generated through hydropower with low production cost and without emissions. This is also one of the main reasons for the rather high share of electricity in some of the heat markets shown in Fig. 5. There are different reasons why there is a widespread use of electricity-based heating in some places but a shift to district heating or other forms of heating where renewable sources are used should be seen as a necessary measure to achieve energy and environmental targets. From a Nordic perspective, where the share of renewables in power production is high, a shift from electricity-based heating would offer tremendous opportunity to meet national and international policy targets through letting the not used electricity displace less environmentally benign power production. The transport sector, now being one of the main carbon-diode emitters, might alternatively benefit from energy-carrier switching to electricity. However, this depends on the maturity and the efficiency of the technology for electric vehicles.

10. Conclusions

District heating is a comprehensive concept for heat supply from fuel through heat production and distribution to consumers. District heating systems are valuable assets, which enable efficient resource utilisation.

The main advantages with district heating are low primary energy demand due to high energy efficiency, high security of supply through utilisation of domestic renewable energy resources, if available, as well as small carbon-dioxide emissions thanks to low fossil fuel use and the high conversion efficiency.

District heating enables utilisation of energy resources that are difficult to use in single buildings and that otherwise may be wasted, such as industrial surplus heat, municipal waste and heat from generation of electricity in combined heat and power plants.

Incineration of waste with heat recovery to district heating may be used at very low cost. Surplus heat from industries can, instead of being wasted to air or water, be utilised in district-heating systems. District heating also gives opportunity for cogeneration of power and heat with high efficiency. District heating, thus, enables profitable heat supply with less environmental impact.

District cooling from, for example, absorption-cooling devices saves electricity and may increase power production in CHP plants. To efficiently utilise biomass for energy purposes, it could be supplied to poly-generation plants where it yields, heat, steam, electricity, cooling and automotive fuel.

The use of, for example, biomass fuel decreases the dependency on imported fossil fuels. Efficient plants need less fuel, which decreases the vulnerability of energy supply. Global warming and better houses reduce heat demand. Using district heating for additional purposes enables increased utilisation of energy resources that otherwise may be wasted.

11. Outlook

Companies and organisations in well-developed district-heating countries have much knowledge that can facilitate district-heating development elsewhere. Such actors could help establishing district-heating systems from fuel supply, via heat production plants and networks to customer contracts. It would promote industrial prosperity for all parties and help building sustainable energy systems in Europe [17].

Government on all levels should recognise district heating as means for increased efficiency of energy utilisation, higher security of supply and decreased environmental impact and their policies should facilitate district heating development.

District heating and cooling can be keys to sustainable local energy systems, which connect energy surplus and energy demand at various temperatures. Regional heating and cooling networks could be supplied by a variety of heat and cooling sources. In such systems, energy supply and demand could be matched, for example, industrial surplus heat, hot water for dishwashers and cooling of rooms, food and water.

European district heating industry has a vision of metering and control of heat sources and consumers that match and optimize energy sources and demand. The vision envisages that IT, real-time smart metering devices and intelligent substations for individual customers, in the future will allow energy inputs and outputs to be identified, matched and regulated in order to optimize the interaction between sources of energy supply and the various temperature demands of customers [24].

Introduction of new district heating systems and modernisation of old ones can result in optimal energy systems from forest to living room.

Author details

Dag Henning[1*] and Alemayehu Gebremedhin[2]

*Address all correspondence to: dag.henning@optensys.se

1 Optensys Energianalys AB, Linköping, Sweden

2 Department of Technology, Economy and Management, Gjøvik University College, Gjøvik, Norway

References

[1] Werner S. The European heat market, Ecoheatcool work package 1. Brussels: Euroheat; 2006. www.euroheat.org/ecoheatcool (accessed December 2007).

[2] Danestig M, Henning D. Efficient heat resource utilisation in energy systems. In: Magnusson FL, Bengtsson OW (ed.) Energy in Europe: Economics, Policy and Strategy. Hauppauge: Nova Science Publishers; 2008.

[3] Swedish District Heating Association. http://www.svenskfjarrvarme.se (accessed June 2012).

[4] Energy in Sweden - facts and figures 2010, ET2010:46.Eskilstuna: Swedish Energy Agency; 2010. www.energimyndigheten.se (accessed June 2012)

[5] Henning D, Danestig M, Holmgren K, Gebremedhin A. Modelling the impact of policy instruments on district heating operations – experiences from Sweden. In: Lectures, 10th International Symposium on District Heating and Cooling, 3-5 September 2006, Hanover, Germany. Frankfurt aM: AGFW-VDEW; 2006. http://www.lsta.lt/files/events/13_henning.pdf

[6] Henning D. MODEST: Model for Optimization of Dynamic Energy Systems with Time dependent components and boundary conditions. In: Karlsson M, Palm J, Widén J.(ed.) Interdisciplinary Energy System Methodology – A compilation of research methods used in the Energy Systems Programme, Arbetsnotat 45. Linköping: Program Energisystem, IEI, Linköpings universitet; 2011. p44-51. Available from http://www.liu.se/energi/publikationer/arbetsnotat?l=sv (accessed March 2012).

[7] Unger T, Ekvall T. Benefits from increased cooperation and energy trade under CO_2 commitments—The Nordic case. Climate Policy 2003; 3(3) 279–294.

[8] Vaillancourt K, Labriet M, Loulou R, Waaub J-P. The role of nuclear energy in long-term climate scenarios: an analysis with the World-TIMES model. Energy Policy 2008; 36(7) 2296-2307.

[9] CO_2 emissions from fossil fuel combustion – highlights. Paris: International Energy Agency; 2011. http://www.iea.org (accessed June 2012)

[10] Klugman S, Karlsson M, Moshfegh B. A Scandinavian chemical wood pulp mill, Part 1 Energy audit aiming at efficiency measures. Applied Energy 2007;84 326–339.

[11] Rydén B., editor. Towards a Sustainable Nordic Energy System. Stockholm: Elforsk; 2010. http://www.nordicenergyperspectives.org (accessed June 2012).

[12] Gebremedhin A, Moshfegh B. Modelling and optimisation of district heating and industrial energy system - An approach to a locally deregulated heat market. The International Journal of Energy Research 2004;28 411-422.

[13] Karlsson M, Gebremedhin A, Klugman S, Henning D, Moshfegh B. Regional energy system optimization – Potential for a regional heat market. Applied Energy 2009; 86(4) 441-451.

[14] Henning D, James-Smith E, Holmboe NM. Nordic electricity supply and demand in a changing climate. Linköping: Optensys Energianalys, Copenhagen: Ea Energianalyse; 2011.

[15] Starfelt F, Daianova L, Yan J, Thorin E, Dotzauer E. The impact of lignocellulosic ethanol yields in polygeneration with district heating – A case study. Applied Energy 2012;92 791–799.

[16] Djuric Ilic D, Dotzauer E, Trygg L. District heating and ethanol production through polygeneration in Stockholm. Applied Energy 2012;91 214–221.

[17] Henning D, Mårdsjö O. Barriers to district heating development in some European countries. In: 12th International Symposium on District Heating and Cooling, ISBN 978-99 49-23-015-0, 5-7 September 2010, Tallinn, Estonia. p223-228.

[18] Henning D, Gebremedhin A. Future biofuel utilisation for small-scale heating and large-scale heat, electricity and automotive fuel production. In: World Bioenergy 2008 Conference & exhibition on biomass for energy: Proceedings poster session, 27-29 May2008, Jönköping, Sweden. Stockholm: Swedish Bioenergy Association; 2008. p24-28.

[19] Directive 2010/31/EU of the European parliament and of the council of 19 May 2010 on the energy performance of buildings, http://eur-lex.europa.eu

[20] Åberg M, Henning D. Optimisation of a Swedish district heating system with reduced heat demand due to energy efficiency measures in residential buildings. Energy Policy 2011; 39(12) 7839-7852.

[21] Åberg M, Widén J, Henning D. Sensitivity of district heating system operation to heat demand reductions and electricity price variations: A Swedish example. Energy 2012 in press.

[22] Henning D. Tillförsel utifrån eller nästan självförsörjande: Energiscenarier för den nya stadsdelen Södra Butängen i Norrköping. Linköping: Optensys Energianalys; 2012.

[23] Henning D, Trygg L. Reduction of Electricity Use in Swedish Industry and its Impact on National Power Supply and European CO_2 Emissions. Energy Policy 2008; 36(7) 2330-2350.

[24] District heating & cooling: A vision towards 2020-2030-2050.Brussels: DHC+technology platform; 2009. www.dhcplus.eu

Chapter 2

Potential and Use of Bioenergy in The Association of Southeast Asian Nations (ASEAN) Countries – A Review

A. Q. Malik

Additional information is available at the end of the chapter

http://dx.doi.org/10.5772/51917

1. Introduction

Bioenergy is considered to be the largest renewable and sustainable energy source of the world's total primary energy supply. At the same time biomass provides fuel for production of 1% of the global electricity generation. It provides 26% of the total primary energy supply and accounts for 87% of the renewable energy supply in Southeast Asia [1]. A very strong community similar to the European Union has emerged consisting of ten member countries: Indonesia, Malaysia, Philippines, Singapore, Thailand, Brunei Darussalam, Vietnam, Lao People's Republic, Myanmar, and Cambodia; and known as the Association of Southeast Asian Nations (ASEAN). Biomass is an important source of energy in these countries and its use is still increasing. The rural population of this region and small industries use it as their energy source. Many countries of this region are among the top producers of agricultural products such as rice, sugar, cane, palm oil, coconut and rubber. The other important biomass resources are the agricultural residues such as bagasse, rice husk, palm oil waste, wood waste, logging wood residues, rice straw, sugar cane trash and coconut shells which accounts for more than 120 million tonnes per year [1]. Bioenergy can be converted into heat, electricity, liquid fuels, such as biodiesel bioethanol, methanol, dimethyl ether (DME), or gaseous biofuels like biogas and hydrogen indicating that it is capable of replacing each type of fossil fuel as well as producing clean energies. Literature reports that ASEAN countries produce $30\,million\,m^3$of wood residue, $19\,million\,tonnes$of rice husk and more than $27\,million\,tonnes$of palm oil residues which can produce approximately $41,\,000\,MW$ of power [2]. A substantial amount of these residues are disposed through open burning and dumping while only a small fraction of it is used as a fuel for heat, electricity generation and household cooking indicating that the use of biomass not only provides alternatives to current energy sources but also eliminates disposal problems associated with generated agricultural residues [1].

The United Nations Framework Convention on Climate Change (UNFCCC) has established an international policy framework for reducing greenhouse gas (GHG) emissions through a programme known as "Clean Development Mechanism" (CDM). A number of such projects have been initiated in ASEAN countries which are beneficial to reduce emission of GHG due to open field burning of forest as well as agricultural residues. With these projects not only the emission of GHG is reduced but more sustainable methodologies in managing natural resources to achieve more efficiency has also been demonstrated.

The objective of this study is to report the potential and the present use of bioenergy in the ASEAN countries focusing on power generation potentials using available biomass resources and the utilisation of CDM projects to achieve energy sustainability.

2. Clean Development Mechanism (CDM)

The United Nation Framework Convention on Climate Change (UNFCCC) established international policy framework for reducing greenhouse gas emission that was adopted at the third Conferences of the Parties (COP-3), the Kyoto Protocol aims to stabilize atmospheric concentrations of greenhouse gases at a level that would prevent dangerous climate change. To make this target achievable and cost effective, provision was given that reduction in GHG could be carried out at any location on the globe because ultimately it has the same effect on the environment. Therefore, it is economically more feasible if developed countries reduce GHG emissions in developing countries rather than at home. This flexible mechanism to reduce GHG emission introduces a new concept known as "the Clean Development Mechanism, (CDM)". The CDM enables developed countries to invest in emission reduction projects in developing countries. It will provide the opportunity to the developing countries to achieve sustainable development and assist developed countries in achieving reduction in GHG in cost effective way [2].

The host country undertaking the CDM projects reduces GHG emission and has the potential to earn carbon credits that can then be traded with a buyer (developed country) providing an additional revenue to finance the project. The introduction of this idea provides new opportunities for developing countries to set up projects that would not be otherwise possible without carbon credits and have the potential to [3]:

- improve local waste management practices (disposal of waste through composting or combustion, landfill gas recovery)
- support the use of renewable energy (e.g. combined heat power production from biomass, biogas, solar, wind)
- encourage energy efficiency initiatives (cogeneration, efficient chillers, energy saving lamps, heat recovery)

http://dx.doi.org/10.5772/51917

- waste to energy (disposal and management of municipal solid waste, agricultural and forest residues)

The host country is directly responsible for assessing the sustainability of CDM projects as per Bonn agreement "The Conference of parties agrees to affirm that it is the host party's prerogative to confirm whether a clean development mechanism project activity assists it in achieving sustainable development" (UNFCCC, 2001). The developing countries of ASEAN community (Cambodia, Lao PDR, Myanmar and Vietnam) are lacking in technical know-how along with non availability of data for assessing the sustainability of proposed CDM projects make it difficult to compute the net reduction in GHG emission on completion of the proposed project. Feasibility studies are carried out by hiring foreign expertise to compete for such projects which is time consuming and usually responsible for delay leading to fewer approved CDM projects for these countries [2]. The priority areas identified by the member of ASEAN nations for CDM projects are tabulated in Table 1.

Cambodia [4]	**Indonesia** [5]	**Lao, PDR** [6]
Wetwaste Biogas-electricity Rice husk/woodwaste gasification Waste management Waste to Energy Agro-forestry	Clean Energy Conservations CHG friendly agriculture and husbandry practices Sustainable waste management GHG mitigation in industries and transportation sectors	Afforestation & reforestation Biomass and biogas Energy efficiency
Malaysia [7]	**Myanmar** [8]	**Philippines** [9]
Energy efficiency and Renewable energy Biogas: POME & animal manure Landfill gas Biomass CHP Biofuels Waste management	Energy crops & biofuels Renewable Energy Afforestation & forest conservation	Waste management Renewable energy Afforestation & reforestation
Singapore [10]	**Thailand** [5]	**Vietnam** [11]
Environmental sustainability Economic sustainability Social sustainability	Biomass and biogas Solar and Wind Biofuels Fuel switching (oil to biofuels) Production process improvement Waste to energy	Energy efficiency, conservation and saving Fuel switching Methane recovery and utilization from waste disposal sites and coal mining Application of renewable energy sources Afforestation & reforestation

Table 1. The ASEAN countries proposed priority areas for CDM projects

The registered CDM projects in different ASEAN countries as of 1st November 2009 are given in Table 2.

Country	Total	Active	Rejected
Cambodia	5	5	0
Indonesia	117	95	23
Lao PDR	2	2	0
Malaysia	167	128	41
Philippines	89	75	15
Singapore	8	8	0
Thailand	119	112	7
Vietnam	93	85	8

Table 2. CDM projects in ASEAN countries (adopted from Status and barriers of CDM projects in ASEAN Countries, UNEP) [3]

These projects concentrate on agriculture, biomass, landfill gas to electricity, biogas from wastewater treatment and biogas from biomass. The developed countries of the region: Indonesia, Malaysia, Philippines and Thailand have a large number of active projects in the pipeline while Cambodia, Lao and Singapore have only a few projects. Brunei Darussalam and Myanmar have no CDM projects because Brunei Darussalam has no designated national authority (DNA) or recently established DNA and Myanmar's previous closed-door international policy made it unfavourable. Recently Japan showed interests to support CDM projects for sustainable development in Myanmar. The development of CDM projects highlights the efforts of the host country to opt renewable energy which are available and its potential yet to be realized.

3. Brunei Darussalam

Brunei Darussalam has an annual waste of 189,000 ton and there are six landfill sites: one in Brunei/Muara, two in Tutong, two in Belait and one in Temburong districts. There is a potential of bioenergy from this solid waste equivalent to $1.3\times10^{5}\,kWh\ year^{-1}$[12]. Schnitzer and Ngoc [13] reported that Brunei with a population of 383,000 persons has a waste generation of $0.66\,kg\,cap^{-1}\,day^{-1}$consisting of 22% paper, 44% food waste, 2% plastic, 5% metals, 4% of glass and 13% others. Research and development projects are underway to study the feasibility for generation of electricity from solid wastes.

4. Cambodia

Cambodia consisting of 21 provinces has 24 isolated diesel power systems located in provincial towns and cities. Per capita consumption is only about $48\,kW\ year^{-1}$ and less than *15%* of households have access to electricity (urban 53.6%, rural 8.6%) and the amount of electricity consumption is as follows: private sector 0.5%, service sector 40%, industrial sector 14%. The supply requirements are projected to increase on average by 12.1% per year, and the peak load is expected to reach to *1,000 MW* in 2020 [19, 21]. 85% of the Cambodian population lives in rural areas and less than 10% of the rural households have access to electricity. Most of energy resources of the urban population are dependent on forest and 98% use woodfuel for cooking [14] and as a result its natural forests have been severely degraded due to logging over the last three decades. Researchers recommend that intervention is needed to ensure a sustainable supply of woodfuel exists in the long term. They are optimistic that increasing woodfuel production, better management of forests and firewood plantations, and introducing non-forested sources such as shrubs for cooking can lead to forest sustainability in Cambodia. Kampong Thom Providence has the highest potential of biomass as an energy source. Top et al. [14] and Top et al. [15] claimed that the potential supply was higher than demand indicating that forest resources and use of woodfuel in this providence are sustainable. Top et al. [16] stated dependence on woodfuel should be decreased by replacing traditional cooking methods with more efficient stove types. Abe et al. [17] studied the potential of rural electrification based on biomass gasification in Cambodia and reported that small scale gasification systems capable of generating electrical power in the range of 4kW would be the most appropriate for rural mini-grid electrification. This study revealed that besides the agricultural residues consisting of rice husks, cashew nut shells and sugarcane which have high energy potential, the proper tree farming and plantation could provide sufficient sustainable sources to supply a biomass gasification system. Koopmans [18] reported that total wood biomass for the year 1994 was 82,022 *kton* and for the year 1990 was16, $900\,ton\,km^{-2}$. However, biomass gasification is economically competitive compared with diesel generation but a comprehensive study to quantify biomass production across multiple rotations and with different species across Cambodia is urgently needed.

Japan Development Institute (JDI) and Kimura Chemical Plants Co., Ltd. based on the request of the Office of the Council of Ministers conducted a study on "Cambodia Bio-energy Development Promotion Project" which was partially supported by Engineering Consulting Firms Association (ECFA), Japan and reported that bioethanol and biodiesel can be developed using cassava and Jatropha, respectively and can be grown in Cambodia without intensive irrigation systems. It was recommended that in order to meet the future target for bioenergy production, Cambodia should expand planting for cassava and Jatropha to a few million hectors each by 2020 targeting to become a net exporter of energy [19]. This study provided a foundation for substantial investments from both local and foreign (Thailand, Malaysia, Koeria, China and Singapore) sources in the development of bioethanol and biodiesel. Almost 5% of the Cambodian national land area is given to private companies for the development of agro-industrial plantations [20]. The Government of Cambodia has been providing special concession scheme to investors to invest in biodiesel production that is mainly focused on Jatropha as feedstock crops. A number of initiatives are still under either planning or implementation stages.

Bioenergy, energy efficiency, waste management, deforestation and forest degradation are the potential areas for CDM projects in Cambodia. There are four approved project on biogas, one on waste/heat gas utilisation and one on biomass and completion of these projects would be able to reduce an annual emission of CO_2 of 204, 308 $t\ year^{-1}$[4]. Literature reports that the country lost 29% of its primary evergreen forests to severe degradation between 2000 and 2005 [21]. A case study is under validation to manage these degraded evergreen primary forests in Cambodia for sustained flow of timber and other ecosystem services that could lead to financial incentives through a carbon payment scheme under global climate change mitigation through reduction emissions from deforestation and forest degradation (REDD-plus) scheme [22].

The aggregate technical potential for electricity generation from biomass consisting of forest products, agricultural crops and residues, municipal waste and sewerage has been estimated using computer simulation techniques at18, 852 *GWh per year*. The findings do not explicitly indicate the provision of efficiency in this analysis. Small scale projects based on simplified technologies are most appropriate as CDM projects for Cambodia. However several CDM projects are in implementation or registration/validation phases but low awareness among policy makers and the private sector, weak institutional capacity, lack of human and technical resources, inappropriate policies and strategies are the major limitations to avail opportunities for carbon trading through CDM [3]. Four out of five active CDM projects on rice husk cogeneration, rubber plantation, improved cookstove and biogas in Cambodia would reduce emission by $4.2\ MT\ CO_{2e}$ over a period on 7-30 years.

5. Indonesia

There is a severe reduction in fossil fuel supplies in Indonesia. The current oil and gas reserves are reported to be approximately 747 million cubic meters (94.7 billion barrels) of oil and *2557* million cubic meters (*90,300* billion cubic feet) of natural gas representing a 13% reduction in supplies which is significant because the demand for fossil fuels has already exceeded the supply capacity of Indonesian oil industry [23]. Among the several alternative renewable energies available for the country to be harnessed, bioenergy has generated considerable interests and shows theoretically an adequate potential to overcome the energy shortage and create a balance between the energy demand and supplies for Indonesia. Bioenergy is renewable and reduces CO_2 emissions when substituted for fossil fuels [23].

Indonesia is an agrarian country and has approximately 90 million hectares of forest indicating that it should concentrate on the development of biomass-based energy programs. At the same time it was ranked among the top ten countries on the globe encountering a net loss of forest area during the era 2000-2005 [23] indicating that bioenergy development projects should be designed in such a way that do not aggravate the loss of forests for sustainability and viability for long term applications. There was approximately *13 million Mg* oven-dry-weight of forest biomass in 2005 [24] another study reports that aboveground forest biomass ranged from $\sim 5000\ to\ 11,\ 000\ million\ Mg$[25]. It is reported that the quantity of wet biomass that can be harvested from 'production forest' and 'other land with tree cover' could be approximated in three ranges which are: *5083 million Mg* (a lower bound), *5410 million Mg* (a moderate bound) and *10,726 million Mg* (an upper bound). The wet biomass was converted to dry weight equivalent using data from the Global Forest Resources Assessment 2005: Indonesian Country Report [26], State of the World's Forest 2005 [24]; and global Forest Resourses Assessment 2005: progress towards Sustainable Forest Management [27]. Suntana et al. [23] reported that if Indonesia converts forest biomass into bio-methanol for electricity generation and as a gasoline substitute then annually 10,063,731 households could be provided with electricity continuously using a *1kW* fuel cell. The results reported are obtained using widely accepted calculation methods due to Vogt et al. [23A] which uses the quantity of biomethonal produced from the annually collected forest biomass and the amount of electricity and transportation fuel that could be substituted by the biomethanol produced from the annually forest materials. With the use of only 5% of forest biomass and converting it to bio-methanol as a gasoline substitute would be equivalent to the total quantity of gasoline consumed in Indonesia during the year 2005. The use of bio-methanol as a substitute for fossil fuel to power vehicles could avoid the emissions of 8.3-34.9% of the total carbon emitted in Indonesia. Timber extraction data from the 1980s reveal that *7.5 million* m^3 *per year* log wastes are generated during harvesting operation that corresponds to about *3.75 million Mg* biomass and is equivalent to collecting biomass materials from 124, 000 $ha\ year^{-1}$ of forest land. 29.5 million litres of bio-methanol can be produce with an efficiency of 25% and could avoid *21.7* Gigagrams *(Gg)* of carbon if it is substitute for natural gas-methanol in fuel cells or *1.97 Gg* of carbon when it is used to supplement gasoline.

Indonesia is the third-largest producer of rice in the world and produced 65,150,764 metric ton in 2010 compared with 64,398,890 metric ton in 2009. Rice bran containing 13.5% oil has a potential for extraction of biodiesel. Gunawan et al. [28] studied rice bran for a potential source of biodiesel production in Indonesia and claimed that 96,000 ton of biodiesel can be obtained from rice bran per year.

Oil palms is another energy crop which were grown on 3.6 Mhectares of land in 2005 and Indonesia is strengthening its production with the increasing worldwide demand for biodiesel derived from oil palms. These trees start bearing fruits approximately 30 months after planting in field and continue to be fertile for a period of 20-30 years ensuring a consistent supply of oil. The estimate for the additional land demands for palm oil plantation in 2020 range from *1 to 28 Mha* in Indonesia that can be met to a large extent by degraded land as well as agricultural management such as implementation of best management practices and earlier replanting with higher yielding plants. Palm oil production has played a major role in land use change in Indonesia [29] and it produces 44% of the world's palm oil as per records for the year 2009. It is predicted that palm oil would be the leading internationally traded edible oil by the year 2012 [30] and the palm oil industry in Indonesia looks forward for high pressure modern power plants to cope with future demand. It is estimated that the residue of palm oil consisting of empty fruit bunch, fiber, shell, wet shell, palm kernel, fronds and trunks has a potential for annual power generation of *5000 GWh* [31]. The transition of energy scenario from fossil fuels to biomass has been underway using existing technologies. In order to make it practically effective requires substantial investments in infrastructure, conversion technologies and in research and development (R&D) for palm oil biomass.

The other feedstocks for biofuel in addition to palm oil, forest biomass and rice bran are crops waste (rubber truck, coconut, sugarcane), waste of food crop products (cassava, sjatropha, sorghum,, soybeans, peanuts, maize, paddy) account for 12.77×10^6tonnes per year and 87.45×10^6tonnes per year, respectively [32]. The crops waste are residues left in field after grain harvest. The Government of Indonesia is in the process of preparing additional land for growing high-yield feedstocks to meet the country's biofuel production goals of 5.57 million kiloliters of biodiesel and 3.77 million kiloliters of bioethanol [33].

The U.S. Department of commerce claimed that biomass installed capacity for energy source in Indonesia is *445MW* which is only 1% of the total resource potential of 49,810MW. The country has targeted 810MW with a conversion efficiency of about 30% of biomass power by 2025 with an increase of 83% but still it is far less than the potential contribution [33].

Research conducted at BPPT-LSDE in Indonesia reported on a plan to construct 1500 litre per day capacity biodiesel using palm oil waste. The domestic manufacturing capacity of biomass gasifier is improved and capable of producing $15-100\,kWe$for rice mill and wood mill power supply as well as for rural electrification. The Indonesian Government has been focusing its policy on bioenergy diversification and introduced a huge plantation of Jatropha curcas as an additional biodiesel source which is non-edible and has well known potential to be converted into biodiesel [34].

Jupesta [35] studied technological changes in the biofuel production system in Indonesia using mathematical modelling consisting of two scenarios: the base scenario and the technology scenario. The base scenario assumes the conditions and data set in the Indonesian Government's Mix Energy policy that relies on an increase in biofuel production by increasing the land allocation for biofuel while the technology scenario concentrates technical change consisting of growth in yield and a cost reduction in addition to the growth in land allocation. The author reported that the highest contribution is likely to come from palm oil that accounts for 93% and 64% of the technology scenario and the base scenario, respectively. The excess production for export increases in both scenarios. But the technology scenario gives more competitive results.

The substantial amount of bagasse in sugar mills can provide fuel for electricity-generating projects in Indonesia that will most probably be considered for the Clean Energy Development Mechanism (CDM) scheme. A recent study concluded that this source has a potential of 260, $253\,MWh$ that could generate a Greenhouse Gas (GHG) emission reduction of 240, 774(large scale) or 198, $177\,tCO_2$(small scale) annually. The present low efficiency cogeneration for those values lead to the earning of about US$1.36 or 1.12 million respectively.

Out of six regional grids in Indonesia where the electricity from the project activities can be grid-connected, primary emission reductions potentials exist in Java, Bali and Southern Sumatera grids [36].

Utilization of palm oil mill effluent (POME) to generate electricity by minimising the emanation of methane gas could reduce GHG emission of 47, 222 tCO_2 per year. The feasibility study of this project that was funded by NEDO Japan is completed and is expected to be considered for CDM scheme and be financed by developed countries. The expected finance for this project could be from Japan [37].

A study financed by the World Bank revealed that the country has a potential to mitigate GHG emissions of over 3 billion tons of carbon dioxide equivalent (CO_{2e}). There are large scale possibilities for emission reductions in the energy sectors. Biomass offers a large potential for CDM projects [3].

6. Lao, P. D. R.

Wood and charcoal were the most dominated traditional energy resources for the period 1996 to 2002 that account for about 75% of the total national energy consumption. Wood is mainly used for cooking and space heating and in rural areas still accounts for up to 90% of the energy consumption. An increase of 4.8% in the total energy use with reference to period 1996-2002 is noted. The Government is keen to develop bioenergy for which more than 2 million hectares of ideal land has been initially identified for biofuels feedstock plantations which is a major step to produce enough biofuels by 2020. Protected area management system is enforced in Lao PDR and a recent study on improvement and implementation of protected area management with positive interaction between people and the natural environment was conducted using a simple simulation model, the "Area Production Model" aiming at evaluating different options for land use and primary production. The findings of this research reveal that the integrated land-use planning approach was found to be well adapted to the needs of the protected area management system [38]. The Ministry of Planning and Investment signed a Memorandum of Understanding in June 2008 with private companies to construct two biodiesel factories with a production capacity of 50,000 tones each by 2010 [39]. A production of the Biodiesel (B100) was reported on July 7, 2011 at a rate of 40,000 Litres/month (*www.linkedin.com/groups/Green-Energy-in-Cambodia-Lao-3991528).* A rural Renewable Energy Initiative in the great Mekong Subregion reports that Lao produces 223,300 tones of sugarcane and 55,500 tones of cassava in the year 2007 indicating that Government policy towards the development of bioenergy is progressing.

A study on "Application of biofuel supply chains for rural development and Lao energy security measurements" was conducted in March 2008 which claims that bioethanol could substitute for 20% of gasoline use in 2030 with the production of commercially viable Jatropha biofuel in four different phases starting from 2008 to 2030 over a total land of 1.1 million hectares [40]. Bush [41] discussed that bioenergy holds enormous potential of 18907 $MWh\ year^{-1}$,equivalent to 1922 $million\ l\ year^{-1}$of diesel fuel in Laos, with an abundance of biomass from agricultural residues (rice husk and livestock manure) and forestry residues (firewood, sawdust, off cuts and woodchips). The author did not elaborate on the conversion efficiency and the heating values which were assumed for these calculations.

A vigorous growth of bamboo is reported in the northern part of Laos that traditionally can be used for construction and handicraft to food and feed. A recent study attributed to bamboo a high potential as a biomass resource for biofuels or fiber, giving a rough review on the potential of three varieties of Japanese origin grown in USA [42]. Northern Laos is considered to be one of the most under-researched regions of the globe and requires further scientific research to investigate bamboo's properties as a biofuel crop.

Lao has only one registered CDM project and another is at its validation stage. The country has a huge potential of CDM projects in forestry but it still has long way to go for capitalizing on the CDM opportunities [3].

7. Malaysia

The palm oil industry is one of the leading industries of Malaysia that produces more than 13 million tonnes of crude palm oil annually resulting pam oil mil effluents (POME) that is approximately three times the quantity of crude palm oil. Wu et al. [43] conducted research on the biotechnological use of POME and reported that in additional to its conversion into useful substitutes for animal feed and fertilizer its fermentation leads to development of antibiotics, bioinsecticides, solvents (acetone-butanol-ethanol: ABE), Ployhydroxyalkanoates (PHA), organic acids, enzymes and hydrogen production. It could also be used as supplementary food in poultry farming. They emphasised that palm industries in Malaysia take appropriate steps to promote cleaner production for POME and their subtle actions could accelerate the research and development for an enhanced POME management. Palm oil is the most suitable and abundantly available feedstock due to its low production cost. The impact of the palm industry on the environment is an important factor which was conducted by Yee et al. [44] by conducting research on life cycle assessment of palm biodiesel consisting of three main phases: agricultural activities, oil milling and transesterification process for production of biodiesel. Comprehensive energy balance and GHG emission assessments were carried out that reveal that exploitation of palm biodiesel could generate an energy yield ratio of 3.53 (output energy/input energy) showing a net positive energy generation that ensures its sustainability. The combustion of palm biodiesel compared to petroleum-derived-diesel cut down the emission of CO_2 by a factor of 38%.

The extraction of palm oil produces a huge amount of biomass from its plantation and milling activities which is much larger compared to other types of biomass in Malaysia. This biomass has a great potential to be converted to either commercial products like animal food, fertilizer or to biofuel and to generate electricity. Shuit et al. [45] studied oil palm biomass as a sustainable energy source for Malaysia and discussed the use of oil palm biomass to bio-based commercial products, synthetic biofuels and for power generation. The researchers highlighted that all conversion technologies discussed in their research are either being used by the commercial sector are still under research and development (R & D). They concluded that with the use of palm biomass Malaysia can become a major renewable energy contributor in the world and become a role model to other countries having huge biomass feedstock.

The Government introduced the "National Biofuel Policy" in 2006 to reduce the huge demand for transport fuel that concentrates on five strategic thrusts: biofuel for transport, biofuel for industry, biofuel technologies, biofuel export and biofuel for a cleaner environment. Early 2006 saw the launch of B5 (Envodiesel) blended diesel with 5% locally refined, bleached and deodorized (RBD) palm olein however this product was abandoned in 2008 when engine manufacturers decided to stop the use of Envodiesel as it clogs the engine in the long run. Therefore, B5 (Envodiesel) palm oil methyl esters with 5% blend of diesel that meets the European Union (EU) standards was targeted for export. For marketing any biodiesel requires certification from the engine manufacturers as fuel as was done in Brazil that uses vast agricultural resources and opted for a different fuel system known as flex-fuel engine to adapt to biothanol (E85). The flex-fuel engines can burn any proportion of blend in the combustion chamber through electronic sensors which sense as soon as fuel is injected and adjust ignition time. However, for biodiesel not a single modified vehicle patent has been developed so far. The researchers suggested that Malaysian car manufacturers look into the improvement in a diesel engine that includes modifications on fuel supply system so that biodiesel developed in Malaysia can also be used in the country [46]. Goh and Lee [47] stated that a palm based biofuel refinery could provide an alternative for Malaysia as a reliable energy supply. With the full use of palm biomass 35.5% of national energy consumption can be secured using a land area of only 8% of current palm cultivation.

A renewable energy feed-in-tariffs (FiT) to support generation of green electricity in the country was introduced by the Malaysian Government under the 10th Malaysian plan which includes all renewable energy technologies, differentiates tariffs by technology, and drives the tariffs based cost of the generation. In the proposal it is also suggested that the FiT programme would add 2% to the average electricity price in the country. Under such a system, electricity generated from renewable energy resources is paid a premium price for delivery to the grid and an exemption for a rise in electricity costs in available for low-income consumers [48]. Chua et al. [49] reported on the feed-in-tariff (FiT) outlook in Malaysia and claimed that this process can lead to a stable investment environment that can generate the development of renewable energy deployment in the country. They quoted the examples of Germany, Spain and Thailand who adopted this process successfully and created more employment, a great investment market and security as it is renewable and helps in reduction of GHG emission. Biomass and biogas including solid waste are expected to be continued as the main sources of renewable energy for the next 20 years. A Municipal solid waste (MSW) of approximately 17,000 tonnes per day throughout the country has been handled by the local authorities and waste management consortia. The largest source of MSW are domestic waste (49%) followed by industrial waste(24%), commercial/institutional (16%), construction (9%) and municipal (2%). It is expected that approximately 9 mil tonnes of MSW will be produced a year by 2020 and the potential of renewable energy generation through waste disposal in Malaysia is extremely high. There are 150 landfill sites in operation that are contributing to the immense potential of land fill gas formation. The Jana landfill Gas power generation plant is connected to the national grid and has a capacity of 2.096 MW using two gas engines rated at capacity of 1048 kW, the landfill receives 3000 tonnes of garbage daily. Malaysia also uses the incineration method for solid waste disposal and has one unit that utilizes 1500 tonnes of MSW per day with an average calorific value of 2200 kcal/kg and daily generates 640 kW of electricity.

The authors claimed that a total of about 2500 *MW* capacity is estimated from *25 mil tonnes* of palm oil residues and 39 $mil\ m^3$ of POME which are used for power generation and cogeneration. In addition to that there is also a substantial amount of unexploited biomass waste from logging, padi, sugar and other residues. It was concluded that the potential cumulative installed capacity of biomass will reach to 1340 *MW* in 2050 from 110 *MW* in 2010. Two demonstration projects under BioGen of total capacity 13.5 MW are under construction and on completion could deliver *10.5 MW* to the national grid. The Biogen project reported that the total potential for biomass and biogas mill waste was *2600 MW per annum* in 2005.

There are three generation of biodiesel feedstock: the first generation is due to food crops (FGC); the second generation deals with non-food crops (SGC); and the third generation extracts it from microalagae and palm oil. Algae are grown in open ponds or photo-bioreactors and can produce in areas unsuitable for agriculture. Due to their high productivity, current yields of algae fuels in test facilities lie well above those of FGC [48, 50]. Goh and Lee [51] reported that total energy potential available from the third generation biodiesel (TGB) is $6.50 \times 10^6\ GJ$ while the energy consumption in transportation section for the year 2007 in the State of Sabah was $7.41 \times 10^6\ GJ$ which is equivalent to 88% of the energy demand of the State. Ahmad et al. [52] presented a comparison of microalgae with other biodiesel feedstocks and concluded that microalgae that grow rapidly with high oil contents have the potential to produce an oil yield that is up to 25 times higher than the yield of traditional biodiesel crops, such as oil palm. The authors recommended that two national car manufacturers: Proton and Perodua should concentrate on the development of flexible fuel vehicles to promote TGB.

ASEAN countries have abundant sources of agro-industrial residues that can serve as feedstocks for production of SGB. Goh et al. [53] studied the potential of SGF in Malaysia and reported that the total capacity and domestic demand of SGB are26, 161 $ton\ day^{-1}$and 6677 $ton\ day^{-1}$, respectively. This indicates that SGB is capable of providing energy to the country provided that lignocellulosic biomass are fully converted into bioethanol and it can reduce 19% of the total CO_2 emissions in Malaysia. The data for the year 2007 on SGB reveal a production of *9549 ktonnes* with predicted increasing waste generation trends in future, the estimated potential of bioethanol is $2.58 \times 10^8\ GJ$ using a net calorific value of 27 GJ per ton. The transport sector of Malaysia consumed total amount of energy equivalent to $6.58 \times 10^7\ GJ$ in the year 2007. The authors determined that if lignocellulosic biomass were fully utilized to produce second-generation bioethanol it would fullfill 35.5% of the country's energy demand. Due to a lack of awareness by the policy makers and other related issues this potential could not be harnessed.

http://dx.doi.org/10.5772/51917

It is noted that future trends of biofuel usage are expected to show an increase in demand, ergo necessitating significant and sustainable sources. Pam Oil may not be able to meet such future production scales as it limits the availability of land for food production, fodder and other crops. Ahmad et al. [54] proposed microgales as a sustainable energy source for bio-diesel production and reported that these are more sustainable source of biofuels in terms of food security and environmental impact compared to palm oil. Micralgae are photosynthetic microorganisms that convert sunlight, water and CO_2 to algal biomass and its total world commercial production is about 10,000 tonnes per year [55] while in a tropical zone its natural production is reported as$1.535\,kg\,m^{-3}\,day^{-1}$.

Singh et al. [56] and Singh et al. [57] discuss the management of biomass residues and urban solid waste respectively. They propose vermicomposing solid organic waste of industrial and municipal origin as a viable alternative technology based on a decomposition process involving the joint action of earthworms and microorganisms, as the end product is pathogen free, odourless, and nutrient rich compared to conventional compost. The application of vermicomposing to agriculture could lead to plant nutrient recycling and the facilitation of soil degradation monitoring. It could also reduce dependency on inorganic fertilizer which is more conducive to a sustainable ecosystem.

There are the technical obstacles related to the development of biofuel in ASEAN countries which are highlighted by Goh and Lee [58]. The feasibility studies should base on fundamental technology and practicalities rather than unrealistic assumptions. It is noted that some countries in this region do not follow pubic tendering for awarding biofuel project and introduced non-professionals in this field creating a fledgling industry which collapse due to deceitful activities. In order to overcome these flaws the policies should be transparent and carefully planned by considering all possible aspects which could be arise in the future. Projects under development should follow up with strict monitoring and the capability of companies involved in biofuel projects should be thoroughly investigated and evaluated prior to issuing licences.

Banana is one of other important crops that are cultivated in Malaysia and the Malaysian climate is most suitable for it. Banana takes almost 10-12 months from planting to harvesting and gives its fruit only once indicating that the crop is to dispose of as soon as fruit-producing period is over leading to a huge source of biomass. However, biomass can directly be converted into energy with direct combustion but the relatively high moisture contents of banana residues suggests that supercritical water gasification and anaerobic conversion would be a better choice and these can give higher conversion efficiency. India is the leading country in this field which converts banana residues into methane using Compact Biogas Plant (CBP) [59]. Tock et al [60] studied the production of banana for the years 2003-2008; and its energy and power potential for Malaysia. They claimed that a total power of $949.65\,MW$ comprising of $80.52\,MW$ from direct combustion and $869.13\,MW$ from anaerobic could be obtained from banana residues which are 5% of the total energy consumption of energy.

8. Myanmar

Energy utilization in Myanmar mainly depends upon traditional energy; 64% from fuel wood, charcoal and biomass; and 35% from crude oil and petroleum, natural gas, coal and lignite and hydropower. 52.5% of the total land area is covered with forest and potential available annual yield of wood-fuel is 19.12 million cubic tons. The cultivation of Jetropha was initiated in 2006 as a national project on 3.15 million acres that will increase to *6 million ha* by 2015, and expected biodiesel production would be 20 million tonnes. Two small scale biodiesel plants were established in Northern Shan State and MICDE (Myanmar Industrial Crops Development) respectively. There are four plants under Government of Myanmar (Ethanol Distillery No. 2 Sugar Mill, Kan-ba-lu Distillery, Taung-sin-aye Distillery and Mat-ta-ya Distillery) in the country producing 667 *tons per day* of 99.5% Ethanol. There are three projects for biodiesel production by the private sector: Technology Company Limited managing 10,000 acres of land at Ayeyarwaddy Division, Ngapudaw TS to cultivate mainly jatropha and later cassava and sugarcane; MICDE is preparing an MOU to carry out biodiesel production with a Korean Company (Hae Joyub Bio Energy Myanmar Corporation) to cultivate 150,000 ha of land provided by MICDE to produce biofuel crops; and Great Wall company is cultivating 1000,000 acres of sugarcane in Northern Shan State to produce bioethanol. There are also Government plans to develop large scale production of bioethanol from cassava and sweet sorghum [61].

The Government of Myanmar is planning to establish biofuel villages at some townships states and divisions where potential biofuel crops can be cultivated. A community-based biodiesel demonstration project is being carried out to educate and introduce the community to the importance of biofuels, their impact on our environment and their economical impacts on the country as a whole and on individuals in particular [61].

The Ministry of Science and Technology is providing services for installing biogas plants designed for small village electrification. There are 105 biogas plants installed generating 945 kW of electricity. There is an estimated paddy production of 22,000,000 tons per year; estimated husk volume 4,392,000 tons per year; and 11,695 (small, medium and large) rice mills. Small and medium scale rice mills use rice husk as fuel to generate steam for steam engines. The rice mills using rice husks for gasification are becoming popular among people. 352,000 tons of husk per year is used to generate electricity [61].

9. Philippines

The Philippines is an agricultural country which generates an average of *36,172.50 tons* of waste annually and the waste generation rate is reported as $0.52\,kg\,cap^{-1}\,day^{-1}$ in urban areas and $0.30\,kg\,cap^{-1}\,day^{-1}$ in rural areas that can significantly contribute the country's energy supply. Apart from agricultural residues woodfuels including woodwastes and fuelwood from forested lands are extensively used. The estimated bioenergy resources from non-plantation biomass consisting of agricultural residues, animal manure, fuelwood released through efficiency improvement of current/base energy systems, fuelwood released through substitution by other fuels, municipal solid waste and back liquor are to be 969 PJ in the year 2010. However, the total bioenergy potential is expected to be rise in future but the consumption of fossil fuel is projected to grow at a faster rate. Literature states that in order to fully utilize the potential of non-plantation biomass concentration should be focus on their development and efficient use [62].

Elauria et al. [63] discussed the total annual biomass production potential from forest in the Philippines is in the range of $3.7-20.37\,Mt$ that can generate an energy of *55.5* to *305.6 million GJ* assuming that energy content of wood is $15\,GJ\,t^{-1}$ and if $1\,Mt$ of woody biomass can generate $1\,TWh$ of electrical power, then the annual electricity generation potential ranges from *3.7* to *20.37 TWh*. It could be concluded that the potential of electricity generated trough bioenergy plantation would lie in the range of 3% to 22% of the country's projected electricity demand for the year 2008 and it can reduce a significant amount of GHG emission. The reported results are based on the theoretical model consisting of three possible schemes: incremental biomass demand (IBD), sustainable biomass demand (SBD), and full biomass demand (FBD).

In February 2004, the Government of Philippines through a Department of Energy Circular made it compulsory for the incorporation of one per cent of coconut biodiesel blend in diesel fuel for use in all government vehicles. The president of Philippines in January 2006 introduced a law "The Biofuels Act 2006" that focused on the future development and use of this fuel in the country initially consisting of 5 per cent proportions for bioehtanol and one per cent for diesel blend with provisions for increasing their blend as recommended by the National Biofuels Board (NBB). The Philippines National Oil Company-Alternative Fuels Corporation (PNOC-AFC) was given a task for "identification and development of low-cost biofuel feedstock: jatropha for biodiesel and sweet sorghum and cellulosic for bioehtanol" and identified the following targets to achieve by 2012: 1,500 hectares of jatropha mega-nurseries cum pilot plantations; 700,000 hectares of biofuel crop plantations; and one million MT biodiesel refineries. Later a special clause in the biofuels act was introduced stating that this act shall not be interpreted as prejudicial to the clean development mechanisms (CDM) projects that cause carbon dioxide and greenhouse gas emission reduction by means of fuel use which encouraged and engaged the interests of biofuel producers to introduce biofuels-CDM projects in country [64].

Coconut is one of the three major agricultural by-products of the Philippines and the feasibility study for coconut as a biodiesil was conducted that concentrate on economics, social, political and environmental issues concludes that coconut has a potential for biodiesel production and the energy required for biodiesel processing (thermal energy and electricity requirement) can be met with its residue consisting of husk (*4.1 million tons per year*), frond (*1.8 million tons per year*) and shell (*4.5 million tons per year*). The reduction in the CO_2 emission was estimated to be in the range $3.70-5.01\times10^6$ of tons per year which is 2.85–3.85% of the Philippines' total CO_2 emission in 2010. The authors claimed that the production of biodiesel could further be increased by improving agricultural yields for coconut through improved irrigation; genetic engineering and other technological advances; conversion of additional non-agricultural land into sustainable energy farms; and utilization of alternative feedstocks such as waste grease [65-66]. The other biomass resources in Philippines include residues from rice, maize and sugarcane which are abundantly grown in the country. An assessment of biomass resources conducted by Asia-Pacific Economic Cooperation based on marginal lands (6, 357 km^2 *equivalent to* 2.3% *of area*) reported that the Philippines has a potential of 1, 793, 000 *tonnes per year* and ethanol potential from marginal lands (0.7 hm^3) is equivalent to 13% of current gasoline consumption [67].

Literature reports an average quantity of rice production per annum in Philippines calculated over a five year period (2002 – 2006) is 14,239 Gg that generates 10,680 Gg of rice straw; 95% of the rice straw is burned in the field and only 5% used for other activities. The rice straw burnt in the field could be used to generate electricity. There are 62 countries in the world currently generating electricity using biomass and this production has steadily increased by an avaeage of 13 TWh per year between 2000 and 2008 [68].

The Philippines currently produces biodiesel from coconut oil and is expanding jatropha production. Ethanol feedstocks used or being considered include sugarcane, corn, cassava, and nipa [69]. Biodiesel production in the year 2007 is reported as 35 ML [70].

10. Singapore

It is reported that total organic waste resources of Singapore in 2006 was 1.91 M tonnes which is 74.4% of the total waste [71]. Singapore uses incineration (waste-to-energy) technology to dispose MSW that involves the combustion and conversion of this waste into energy. This technology reduces the volume of solid wastes by 80-90% making it popular in countries having limited territory for landfills. Four incineration plants in Singapore are Ulu Pandan, Tuas, Senoko and Tuas South with turbine capacity of 16 MW, 46 MW, 56 MW and 80 MW, respectively; these plants generates 980 *million kWh* of electricity per year which is 2-3% of total electricity demand of the country; and 22, 800 $tonnes\ year^{-1}$ of scrap metal is recovered for recycling. The proportions of food waste input treated by the four plants are reported as 12.88%, 16.52%, 34.66% and 39.95%, respectively. A typical incineration plant requires the energy input 70 $kWh\ ton^{-1}$ of waste and generates around 20% ash [72-73].

Neste Oil announced in November 2007 the construction of a biorefinery capable of producing NExBTL renewable fuel with a capacity of *800,000 metric tonnes per annum* and the proposed plant would be the largest renewable fuel refinery in the world. Nexsol (joint venture between Peter Cremer and Kulim group) in joining with Continental Bioenergy and Natural fuels invest in Singapore for biofuels production. There will be a capacity of *1,650,000 metric tonnes per annum* of biofuels in the country with the completion of these two projects [70].

11. Thailand

Thailand is abundant in agricultural residues: rice husk, sugar cane bagasse, wood, cassava, maize, cotton, soyabean, sorghum, caster and palm oil; and the country has a potential of 11.2$TW\ h\ yr^{-1}$ or 2.98 GW of power generation capacity. Sajjakulnukit et al. [74] studied the sustainable energy potential of following biomass resources in Thailand: agricultural residues, animal manure, fuelwood saving potential through improvement of efficiency, flelwood saving potential through substitution by other fuels, municipal waste and industrial waste water. A comprehensive estimation on individual resources was conducted using data of harvested land and production statistics from the Centre for Agriculture Information (CAI). They claimed that the total energy potential of these sources is expected to be *821 PJ* for the year 2010 that corresponds to 14% of the total primary energy consumption in the same year [74].

The biomass is processed to generate either electricity or heat using conventional power plants. For successful utilization of biomass for energy production a continuous and secure supply of it to the power plants is the fundamental requirement. Sometimes biomass projects could face difficulties due to limited accessibility, logistical problems, seasonal availability, variation in biomass prices and increased utilization for other applications. Junginger et al. [75] described a methodology to set up fuel supply strategies for large-scale biomass conversion units (between *10* and $40\ MW_e$). The proposed methodology was demonstrated on a case study in an agricultural region in Northeastern Thailand. The study examined variations in residue quantities produced, limited accessibility of residues, utilization by other competitors and logistical risks. Four expected major risks in near future were considered. The first is an increased demand for residues as fuel, especially rice husk; the second risk is the possibility of a bad harvest; the third is the possibly increased demand of rice straw and sugarcane tops and leaves as a raw material for the pulp and paper industry; and the fourth concerns transportation and logistics. To overcome these risks different fuel supply scenarios were incorporated to show how biomass quantities and prices may vary under different conditions. It was noted that both residue quantities and prices can vary strongly which are dependent on fluctuating harvests, increased utilization by competitors and varying transportation costs. The researchers concluded that the combustion biomass plant is economical for agricultural residues.

The amount of agricultural residues (paddy, sugarcane and oil palm) estimated in the year 1997 was about *61 million ton*, of which *41 million ton* equivalent to about *426 PJ* of energy, was unused. The potential of biogas resources due to industrial waterwaste and live stock manure is 7800 and 13, 000$TJ\ yr^{-1}$, respectively. Prasertsan and Sajjakulnukit [76] identified that one of the barriers to promote biomass energy production projects in Thailand is the lack of awareness and confidence that created misconceptions in the Thai people on the use of renewable energy in general and biomass energy in particular. This is because of the fact that education on acid rain that can destroy crops produced a negative impact on the common man and they consider a biomass power plant as a monster. They stressed the need of a new policy approach to overcome the barriers for utilization of biomass energy in Thailand.

Krukanont and Prasertsan [77] used Geographic Information System (GIS) to compute the potential of rubber for power generation in the rubber dominated growing area of south Thailand. The authors identified the location of eight potential rubber-fired power plants along *700 km* of the highway in this region which are economically feasible with a total capacity of 186.5 *MWe* having the fuel procurement area in the range of less than *35 km*.

The Government of Thailand encourages the production of bioethanol from local energy crops like cassava and sugarcane to prevent energy risks when crude oil prices fluctuate greatly or increase rapidly. This fuel has been used for vehicle in various types of ethanol blend with gasoline (known as gasohol) i.e. E10, E20 and E85. E10, a 10% blend of bioethanol with 90% gasoline, was introduced in the market in 2004. E20, a 20% ethanol blend, was introduced in 2008 after E10 had penetrated the market. Later, E85 gasohol was launched in 2008 [78]. Due to Government promotion strategies, the total gasohol consumption in Thailand has increased from $0.61 Mlday^{-1}$ in 2004 to $10.48 Mlday^{-1}$ in 2008 [79]. The Thai renewable energy policy has set the target to increase the use of ethanol up to $3 Mlday^{-1}$ by 2011. The growing demand for biofuels increases the demand for feedstocks that in turn is anticipated to increase the land for agriculture requirements. To cope with the target for 2011 about $6.36M\ ton\ yr^{-1}$ of cassava are needed with an assumption of *133.21* ethanol produced per ton of cassava roots. This demand will change the cropping system and land requirement which contribute approximately 58-60% of the net green house gas (GHG) emissions. GHG emission can be reduced by increasing productivity rather than cultivating land [78]. Silalertruksa and Gheewala [80] stated that feedstock efficiency would be increased and hence GHG emission be reduced by improving soil quality with organic fertilizers, preventing the sugarcane leaves burning during harvesting, enhancing the waste recycling program from ethanol plants such as biogas recovery, organic fertilizer and DDG or DDGS production. They recommended that the use of renewable fuels in ethanol plants, implementing energy conservation measures and providing technical knowledge associated with cassava ethanol production to the industry also need to be encouraged.

Fluidized bed technology is used to convert agricultural and wood residues into energy and emission of various pollutants from this process depends on fuel analysis, combustor design and operating conditions. It is reported that unburned pollutants are expected at insignificant levels with supply of sufficient combustion air. Permchart and Kouprianov [81] studied combustion of three different biomass fuels: sawdust, rice husk and pre-dried sugarcane bagasse in a single fluidized combustor (FBC) with a conical bed using silica sand as the inert bed material and fairly uniform axial temperature. They reported that emission of CO for rice husk was much greater than those for sawdust and bagasse for similar operating conditions due to the presence of coarser particles and higher ash concentration. They noted that the emission of CO can be rapidly diminished with an increase in excess air of up to 50—60% but has a weak dependence with excess air in the range of 60—100%. The emission of NO strongly depends on the fuel-nitrogen contents rather than operating conditions. They concluded that sawdust is the most environmentally friendly biomass whereas rice husk produce noticeable environmental impact. A maximum efficiency of 99% was obtained for sawdust and bagasse at the maximum combustor load with an excess air of 50-100%. The maximum combustion efficiency for rice husk with excess air of about 60% was 86%. A further increase in excess air for rice husk decreases combustion efficiency rather than an increase compared with those of sawdust and bagasse.

Janvijitsakul and Kuprianov [82] investigated the emission of CO and NO_x in a newly-built, $400\,kW$ conical fluidized-bed combustor for firing $80\,kg\,h^{-1}$ rice husk with medium-ash contents and high-calorific value $(LHV = 15.7\,MJ\,kg^{-1})$ for wide range of excess air, 3-59%. They reported the emission of NO_x as 75-143 ppm (corrected to 7% O_2 dry flue gas) at elevated bed temperature of $900-950\,^oC$. The emission of CO was 128-716 ppm (7% O_2 dry flue gas) which shows an inverse correlation with excess air. The total Polycyclic Aromatic Hydrocarbons (PAHs) emission was found to predominate for the coarse ash particles due to the effects of a highly developed internal surface in a particle volume. The highest emission was noted for acenaphthylene, $4.1\,\mu g\,kW^{-1}h^{-1}$ at the total yield of PAHs due to fly ash was $10\,\mu g\,kW^{-1}h^{-1}$.

Shrestha et al. [83] examined the development of energy system during 2000-2050 and its environmental implications in Thailand. The energy system was ranked into two components: energy supply and conversion (energy extraction, imports and conversion of primary energy to secondary energy i.e. power generation), and service demand. New as well as twenty existing technologies were considered for power generation in four different scenarios: *global market integration* (TA1); *dual track* (TA2); *sufficiency economy* (TB1); and *local stewardship* (TB2). They concluded that industry and commercial sectors remain the most intensive user of electricity throughout the study period. The share of coal and natural gas for power generation would account for almost 85% while 15% was from renewable energies like biomass and hydro. Energy used by road transportation considerably increased during this period. In the last two decades of study period clean fuels based vehicles become important and play a dominant role. Thailand imports energy which would increase from 50% in 2000 to 89% in 2050 and this could impose energy security issue for the country. A considerable increase in emission of SO_2 and NO_x was estimated with reference to that of the emission in 2000. In all four scenarios, SO_2 emission would be much faster that the NO_x emission which is due to the substantial use of coal for electricity generation.

The utilization of rice straw residues for power production can improve the renewable energy development plan in Thailand. The major goal for biomass-fuelled power plants is to deliver energy at a reasonable cost. Delivand et al. [84] conducted economic feasibility assessment of rice straw utilization for electricity generation through combustion in Thailand and concluded that to ensure a secure fuel supply smaller scale power plants with capacities $5-10\,MW_e$ are more practicable. Literature reports the production of biofuels in the year 2008 of 822 ML [70].

The Ministry of Energy (MOE) of Thailand stated three main sources of biomass namely agricultural residues, forest industry and the residential sector. The potential and targeted capacity of biomass is *1751 MW* corresponding to *623 MW* from rice husk, *106 MW* from bagasse and *32 MW* from wood residue. The Energy Policy and Planning Office (EPPO) disclosed in March 2010 the existence of *76* biomass power plants generating *673 MW* of electrical power, negotiation with the Metropolitan Electrical Authority (MEA) and the Provincial Electrical Authority (PEA) for *30* plants of capacity of *290 MW* are under way, *40* approved plants with a capacity of *290 MW* are waiting for signing Power Purchase Agreement (PPA) contracts and *211* power plants of capacity *1585 MW* are under construction and waiting for Commercial Operation Date (COD) [85]. The Government of Thailand under the 15 years of the Alternative Energy Development Plan (AEDP) lay down targets to generate electricity utilising biomass in 2022 in three phases: short term (*2008-2011*) to achieve generation of power at *2800 MW*, mid-term (*2012-2016*) to attain a power at *3220 MW*, and long-term (*2017-2020*) to reach the objective of generation of electricity at *3700 MW*, respectively [85]. The 15 years of the Alternative Energy Development Plan (AEDP) presents electricity generation utilising biogas in 2022 in three phases: short term (*2008-2011*) to achieve generation of power at *60 MW*, mid-term (*2012-2016*) to attain a power at *90 MW*, and long-term (*2017-2020*) to reach the objective of generation of electricity at 12*0 MW*, respectively [85]. Similarly for MSW: short term (*2008-2011*) to achieve generation of power at 78 *MW*, mid-term (*2012-2016*) to attain a power at 13*0 MW*, and long-term (*2017-2020*) to reach the objective of generation of electricity at 16*0 MW*, respectively [85]. It is reported that to date 1,610 MW, 46 MW and 5 MW of electricity generation has been obtained using biomass, biogas and MSW.

Literature reports industrial wastewater and livestock manure are the major resources of biogas in Thailand that have a potential of *7800* and *13,000 TJ per year*, respectively. This amount of waste can produce *620 million* m^3 of biogas. The installed capacity of biogas power is *146 MW*. The Energy Policy and Planning Office (EPPO) highlighted in March 2010 the working of *41* biogas power plants feeding *43 MW* of power to the grid, negotiations are in progress with the Metropolitan Electrical Authority (MEA) and the Provincial Electrical Authority (PEA) for 15 plants with a capacity of *41 MW, 31* approved plants with a capacity of *44 MW* are waiting for signing Power Purchase Agreement (PPA) contracts and *33* power plants with a capacity of *72 MW* are under construction and waiting for COD [85]. The Government of Thailand under the 15 years of Alternatives Energy Development Plan (AEDP) lay down targets to generate electricity utilising biogas in 2022 in three phases: short term (*2008-2011*) to achieve generation of power at *60 MW*, mid-term (*2012-2016*) to attain a power at *90 MW*, and long-term (*2017-2020*) to reach the objective of generation of electricity at *120 MW*, respectively [85].

Thailand generates approximately *14.5 million tonnes* of MSW annually consisting of food waste (*41-61%*), paper (*4-25%*) and plastic (*3.6-28%*) which is decomposed to produce landfill gas comprised of *60%* methane and *40%* CO_2 using 90 landfills and three incinerators. The installed capacity of generating electricity from MSW is *13 MW*. The Energy Policy and Planning Office (EPPO) declared in March 2010 that 8 MSW power plants are in operation generating 11 MW of electricity which is fed to the grid, negotiations with Metropolitan Electrical Authority (MEA) and Provincial Electrical Authority (PEA) are being held for *10* plants with a capacity of *305 MW, 15* approved plants with a capacity of *68 MW* are waiting for signing PPA contract and *14* plants with of capacity *96 MW* are under construction and waiting for COD [85]. The Government of Thailand under the 15 years of AEDP lay down targets to generate electricity utilising biogas in 2022 in three phases: short term (*2008-2011*) to achieve generation of power at *78 MW*, mid-term (*2012-2016*) to attain a power at *130 MW*, and long-term (*2017-2020*) to reach the objective of generation of electricity at *160 MW*, respectively [85].

The contribution of Stainable Development (SD) of a CDM project is interpreted by the host country, which develop their own SD criteria for assessing CDM projects. There are no common international standards for the host country approval processes and the development of SD criteria. Stakeholder preferences towards the SD of CDM projects are not explicit and are left to the host countries to interpret. Kerr and Parnphumeesup [2] carried out research using quantitative and qualitative methods to investigate stakeholder preferences towards SD priorities in CDM projects. This study investigate CDM's contribution to SD in the context of biomass by taking a rice husk project as a case study conducted in Thailand. Their quantitative analysis demonstrated the use of renewable energy as a highest priority followed by employment and technology transfer. Qualitative results obtained from this project revealed that rice husk CDM projects could contribute a lot to SD towards generation of employment, increase in the usage of renewable energy and transfer of knowledge but it definitely produces a potential negative impact on air quality. Stakeholders advised that in order to ensure the environmental sustainability of CDM projects Thailand should cancel an Environmental Impact Assessment (EIA) exemption for CDM projects with an installed capacity below 10 *MW* and apply it to all CDM projects. They recommended that the Government of Thailand develop a biomass commodity market to highlight the importance of rice husks for the country's renewable energy plan. Alternatively, farmers form cooperatives that could enforce mills to buy paddy rice at higher price making sure that the true value of the rice husk is paid.

12. Vietnam

Biomass resource potential on marginal lands which is 6.5% of the total area of the country is reported to be 11, 281, 000 $tonnes\ year^{-1}$ and ethanol potential from this land at a rate of 4.4 $h\,m^{3}$ is *79%* of the current gasoline consumption [86]. Biofuels development in Vietnam is in its early stages compared with other ASEAN community, biofuels plants are in process of cultivation, potential feedstock are cassava, sugarcane, rubber seeds, jathropa and catfish oil. The country has a strong national target for biofules as alternative to fossil fuel and woking on ethanol and vegitable oil to replace 1% of country's petroleum demand by 2015, and 5% in 2025.

The biomass resources in Vietnam are: agricultural (paddy, maize, cassava, sweet potato), forest (natural, planted, wood, dispersed), industrial crops (sugarcane, peanut, coconut, cotton, jute, sedge, elephant grass) and other waste (industrial residues consisting of sawdust and molasses, livestock residues and solid waste) which accounts for 60-65% of the primary energy consumption and is being used for cooking fuel, organic fertilizer, biogas for domestic cooking, electricity production (in paper mills) and bioethanol production. The Government of Vietnam introduced a state biofuel development program in November 2007 aiming to develop renewable biofuels from biologically derived organic resources to replace a part of fossil fuels for future State energy security and environmental protection. The targets for these programs are: to develop *100 thousand tons* of E5 and *50 thousand tons* of B5 (0.4% of mass fuel consumption of the country) by the year of 2010; 250,000 tons of ethanol and vegetable oils equivalent 5 *mill tons*of E5 and B5 by the year 2015; and the production of ethanol and vegetable oil will reach 1.8 *mill tons*by the year 2020. A joint project between Petrosetco Vietnam and Itochu Company Japan is constructing a bioethanol factory that would be on completion capable of producing 100 *mill liters per year*using cassava starch. The country is in the process of developing new types of biomass as raw materials for biofuels from sea known as "Kappaphycus alvarezii (Green and Brown), Gracilaria tenuistipitata". The literature does not report any production current activities of vegitable oil in Veitnam. Projects are largely still in the developing stage under the government, save for B5 production levels sourced from fish operating at 50,000 tonnes per year [86A, 86B].

The main feedstocks for biodiesel production in Vietnam are "Basa" fish oil, used cooking oil and rubber seed oil. The potential of "Basa" for the year 2005 was estimated to be of *60,000 tones* that could produce *48,000 tons* of biodiesel. Saigon Petro and Agifish are in the processing of developing a project with a capacity of producing 10, 000 *tons per year*biodisel. There is *73,800 tons* of used oil with a potential of producing 33,000 tons of biodiesel. Saigon Petro is developing a facility to extract 2 *tons per day* biodiesel using 4−5 *tons per day* of used cooking oil. Rubber trees are planted on more than 500, 000 *ha*in 2006 and it is planned to rise to 1, 000, 000 *ha* that can produce 200, 000−300, 000 *tons* rubber fruits every year, equivalent to 17, 600−330, 000 *tons*of rubber oil which is not edible and one of important bio-resource to biodiesel production. It can either be used directly by thermal cracking to hydrocarbon or in form of ethyl ester, blended with petroleum diesel. Biodiesel from ethyl ester of Vietnamese rubber seed oil according to the European standard for determination of biodiesel (E.DIN 51606), blended from 5% to 20% with petroleum diesel, can be used as a fuel for electric generators and car diesel motors [87].

Many projects have been carried out to develop cultivation of jatropha in different providences of Vietnam and some of these projects are in pilot scale for production of biodiesel [87]. Nguyen [88] studied the rice husk potential of Vietnam (1995 – 2002) and noted a rise in the planted area (6,766,000 to 7,485,000 hectors); rice husk output which was assumed to be 20% of paddy increased from 4,993,000 tons to 6,813,000 tons; 30% of the rice husk was assumed to be used to generate electricity and with this assumption a rise in the supply of rice husk for generating power was increased from 923,000 tons to 1,249,000 tons. This study reports that there are 615 rice mills in the country and each mill collects rice within a radius of 20 km. The electricity generated using rice husk that feeds power to a grid can reduce the emission by 0.615 $kg\ of\ CO_2\ per\ kWh$ compared with the conventional fossil fuel. Off grid can reduce 0.8 $kg\ of\ CO_2\ per\ kWh$ [89].

Rice husk and straw are the most available biomass for energy production in the Mekong Delta. Dang et al. [90] studied energy needs for this region by estimating the current and future energy demands of rural industries; identifying the type and quantity of most availa-

http://dx.doi.org/10.5772/51917

ble source of energy production; and developing and assessing biomass utilization scenarios assuming various system scales and conversion technologies. Their findings reveal three important facts. Firstly, electricity and heat energy obtained from rice husk burning furnaces, kilns, or stoves are the energy sources highly in demand by Mekong Delta's rural industries in both the present and the future. Secondly, the biomass based power plants use rice husk and straw as a fuel for generating power which accounts for 73-87% and 8-10%, respectively, of total unused agricultural residues in 2007-2030. Thirdly, the use of biomass power plant due in 2007-2030 could potentially reduce emission by $163-871\,kTCO_{2-eq}$, equivalent to $21-109\%$ of GHG emission in the study area over the same period of time.

Literature reports that a 4.4 GW of renewable energy moderate capacity potential exists in Vietnam that could be utilised to replace conventional fuel-generating capacities to produce electricity in the country. Among other renewables like hydro energy and geothermal energy, biomass resources consisting of rice husk, paddy straw, bagasse (sugarcane, coffee husk and coconut shells), wood and plant residues have a potential of $1000-1600\,MW$ generating electricity equivalent to 22.7% of the total expected potential of renewable energy for the country that accounts for $4.4\,GW$ [91].

The largest obstacle to implementation of CDM projects is lack of technical knowhow, difficulties calculating emission reductions and submitting the requisite evidence of 'additionality' as compared with the business-as-usual scenario. The energy sector is also faced with a lack of reliable official data on the Vietnamese power grid, making it more difficult to calculate viable emission factors and baselines for ascertaining CO_2 savings.

13. Unified ASEAN bioenergy outlook

ASAEN countries are the main producers of palm oil and rubber with substantial plantations of coconut and paddy fields and they have started cultivation of jatropha on large areas. ASEAN countries are located in the equatorial region of the globe that provides a constant warm temperature and humid conditions throughout the year and makes this region suitable for a variety of large areas of plantation. The region has a potential of unwanted biomass (wastes from only palm oil, sugarcane and rice excluding all others) of up to 208.68 *million tonnes per annum* that is generated from the by-products after the milling process. This unwanted biomass has a potential of generating electricity up to 71.47 *TWh* which was 14.37% of total residential electricity usage for all ASEAN countries in 2006; with a conversion efficiency of 30% and $10\,MJ\,kg^{-1}$biomass energy yield. There is a huge agricultural land available in Mynmar, Vietnam, Cambodia and Lao that is still unused due to lack of sufficient funding, infrastructure and a skilled workforce. At the moment these countries are considered as undeveloped and use traditional agricultural methods for cultivation. Under the ASEAN cooperation framework, these countries can take help from other members through technology transfer and skilled manpower to modernise their agricultural sector and increase agricultural revenues. It is expected that with this cooperation the total land area of 10.74 *million hectares*, after deducting the area of plantations from total agricultural land for Myanmar, Vietnam, Cambodia and Lao, can be increased. If oil palm which is one of the highest yield crops is cultivated, assuming that this land is suitable for it, 664.40 *million tonnes* of extra biomass residues can be produced annually which would generate electrical power of 220.68 *TWh* with the same assumptions as stated above. 58.74% of residential electricity usage including the earlier estimate can be generated from biomass leading to a huge reduction in carbon emission [74]. The global contribution of Asia in biofuel production is 4.6% and the ASEAN share lies in the range of less than 2% [74A].

It is noted that highly urbanized cities in ASEAN countries generate a high percentage of organic and mixed inorganic waste (55-77%), with about 10-16% made of plastic, approximately 4-10% of glass and about 4-12% of metal. The largest fraction of MSW in ASEAN countries is paper and cardboards constituting 28% of the waste. There is about 529, 500 *tons per day*urban MSW in ASEAN countries. Out of which approximately 50-80% is collected each day and then disposed in landfills or dumpsites. The capacity of landfills is mostly exceeded due to lack of waste management planning. In countryside waste is either thrown directly into rivers, dumped at the road side or burned in the open because of a lack of finance, land acquisition problems, lack of awareness of the environment, inadequate solid waste management, and lack of enforcement that could impose serious environmental pollution problems [74]. The amount of MSW dumped openly and/or burn is not known.

At the same time one of the most sophisticated waste treatment systems, "incineration" has successfully been used in Singapore. Malaysia has one municipal incinerator and planning for another one, Indonesia and Thailand also have one in their capital cities. Recycling is also becoming popular in this part of the world. In high income countries like Singapore approximately 44.4% of solid waste is recycled. In the middle income countries, the percentage of waste recycled is about 12%, and it is approximately 8-11% for the rest of ASEAN [74]. If 529, 500 *tons per day* urban MSW in ASEAN countries is wisely and professionally treated as Singapore does, it has a potential of generating 271 *GWh per day* of electricity. Research and development (R &D) projects in Malaysia, Thailand and Indonesia are investigate vermicomposing of solid organic waste from industrial as well as municipal origin [56].

Shi et al. [92] estimated the global potential of cellulosic ethanol from waste paper and cardboard to be *82.9 billion litres* and reported that the substitution of gasoline use with waste paper-derived cellulosic ethanol could offer GHG saving of between 29.2% to 86.1% [92]. The introduction of a proper MSW management system in the ASEAN community could lead to a clean environment and it is this area where CDM projects can play a prominent role for sustainable development by reducing the emission of GHG of this region.

A study conducted on construction waste generation and management in Thailand claimed that on average 1.1 million tons of construction waste is generated per year and if the management of this waste material is given attention by prompting recycling an average saving of 3.0×10^5 *GJ per year* could be made [93]. Researchers focused on the introduction of proper MSW management and disposal systems along with strong Government commitment in ASEAN countries which has a high potential of generating electricity from this unavoidable and ever increasing source.

http://dx.doi.org/10.5772/51917

It is desirable that in ASEAN countries waste treatment facilities should be strictly regulated and protected regarding licensing, authorization and compliance with the country's law. Enforcement of law to ensure the regulatory framework must be applied strictly and if necessary existing law on waste management be amended or new laws introduced to protect and minimise environmental pollution through open burning of any type of waste including agricultural, forest, MSW, industrial liquid waste discharges and gas exhaust. The region should concentrate on and opt for available "waste to energy" technologies to deal with all types of these wastes like agro-based industrial systems, recycling, bio-digestion, bio refineries, bio-extraction etc. Developed countries like Malaysia and Singapore help other developed and developing countries of this community to enhance and introduce sustainable waste management system through joint R & D projects and sharing their resources [74].

Effective utilization of biomass as an energy resource is based on biomass availability, transportation distances, and the scales and locations of power mills/factories within a region. Palm oil mills use small boilers for both electricity generation and palm oil extraction processes. The most common type of power plant used in ASEAN countries consists of a small tube boiler capable of processing 30-60 tonnes of full fruit bunches (FFB) per hour that can produce an excess heat and electricity of $23.8\,MJ\ per\ ton\ FFB$and $22.4\,MJ\ per\ ton\ FFB$, respectively. These conventional boilers should be replaced with high pressure boilers such as dual fire boilers capable of burning palm oil waste as well as use of use of POME derived biogas as a supplementary fuel for efficient production of power and heat from biomass. Energy efficiency could also be improved by the adoption of high efficiency motors, high efficiency transformers and variable-speed controls in power plants. Literature states that $2-8\,MW$or $12\,MW$combined heat and power (CHP) plants are most appropriate and can generate the largest profits in Malaysia as well as throughout the ASEAN countries. Installation of large plants requires Empty Fruit Bunches (EFB) transportation over longer distances and couples with low stability of EFB supply particularly in low season. It is recommended that small power plants are installed in such a way that each power plant has a collection area for FEB within approximately a radius of 40-50 kilometres which will ensure the supply of FEB to plants and protect it if left to decay on-site due to limitations of either a plant capacity or difficulties on transportation over long distances for large plants [43, 56]. Malaysia, Indonesia and Thailand among the ASEAN countries have huge resources of biomass from the palm oil industries and these countries can help developing countries of this community to build up biomass conversion technologies by providing expertise as well as skilled manpower. With the close cooperation within the ASEAN community bioenergy technologies are able to penetrate resulting in production of biofuels, generation of electricity using wasted agricultural and other types of residues in the region that can compete with conventional fossil fuels more economically that could lead to sustainable energy development [94].

Forests in ASEAN are an important source of timer and other forest products, of energy for cooking and space heating for the rural population and a potential source of bioenergy. Literature reports that these forests produce about $563.8\,million\ tons\ per\ year$ ($11.3\,EJ$)of woody biomass for the period 1990-2020 and a decrease of 1.5% in annual woody biomass was noted for this period. It was highlighted that if this trend is continued then this region of the world could face a shortage of woody biomass as well as its ecosystem functioning can be adversely affected leading to the risk of its sustainable development. The region has strongly reacted and enforced laws against deforestation and forest degradation. Rehabilitation and plantation programs have been initiated resulting in the recovery of about 0.1% of the 2.4 million ha deforested land. It is claimed that use of woody biomass to replace fossil fuel for energy generation could prevent carbon emission of about 169.0−281.7 *TgC per year*, where one tetragram carbon (TgC) is one million tons of carbon, between 1990 and 2020 [95]. A case study of household energy demand of a rural community and its electrification in Lao People's Democratic Republic was conducted. Prior to electrification 99% of the primary energy demand was met with firewood. Only 75% of villages used commercial lighting fuels while 25% have no access to this fuel and therefore are not engaged with entrepreneurial activities. These families were wasting thousands of hours of productive time each year which could be used to improve their families' living conditions through education and safer time-saving work if they could have access of about three hours of lighting per day indicating the importance of energy and its impact on the lives of rural people. A proper management of forests could solve these very simple problems of the rural communities and enhance their productivity [96].

Biofuels are growing steadily in ASEAN countries which are extracted from sugarcane and cassava; 75% of the current biethanol production in Thailand is from cassava. Thailand is the largest producer of bioethanol with517 *Ml per year*, followed by Philippines (116 *Ml per year*) Indonesia (77 *Ml per year*) and Singapore (34 *Ml per year*). The international energy agency (IEA) stated that Thailand will continue to lead ethanol production in ASEAN countries for the next three years to achieve an expected annual production of 1276 *Ml* in 2012 while Indonesia and Philippines production will further increase to 355 *Ml* and 332 *Ml*, respectively in 2012. Malaysia and Vietnam do not produce ethanol for use in the transport sector [48 & 97]. Palm oil is the main feedstock for biodiesel production in this region of the world with an average biodiesel yield in the range of 4, 000−4, 700 *l per ha* (FAO 2008). Literature states that Thailand led in the production of biodiesel in the year 2009 with a quantity of 625 *Ml* while Indonesia, Malaysia and the Philippines produced 243 *Ml*, 203 *Ml* and 96 *Ml* respectively, whereas Singapore had a comparatively small output of 48 *Ml*. A steady increase in the production of biodiesel for the next three years is predicted led by Thailand with an annual output of 955 *Ml* in 2012, followed by Singapore with an expected increase in production by almost twentyfold to 946 *Ml per year*. This increase is due to a large plant which is currently under construction capable of increasing the country's capacity by900 *Ml per year*. The expected increase of biodiesel in Malaysia, Indonesia and Philippines lies in the range of *25-60%* [49, 97].

There is a huge potential for increasing the power efficiency of energy plants. This can be done by increasing steam parameters and installed power in cogeneration plants and reducing consumption in process. Biogas can be generated from the anaerobic treatment of the liquid effluents of the process and its conversion into electricity using internal combustion engines or micro-turbines. The extraction methodologies used to extract cooking oil and biofuels from biomass could be modified to increase the efficiency [98]. This can be done with use of suitable catalysts to catalyze the transesterification reaction for extraction of cooking oil/biofuels from biomass. Edric et al. [99-100] claimed that conversion of biomass into biofuel/cooking oil and apparent bulk reaction rate are insensitive to temperature but dependent on mass transfer rate and their results reveal that overall reactor performance may be further improved by increasing the porosity of the biomass. It is desirable to investigate further how to improve the catalyst and elucidate the reaction mechanism to increase the quality of biofuels extracted from biomass.

A lot of interest in investing in biomass power in ASEAN countries especially in Malaysia and Thailand has been reported under carbon finance opportunities through CDM projects. It has been reported that among the registered CDM projects for ASEAN countries 41% are on biomass power generation. The majority of the registered CDM projects are small scale under 10MW that are often located in remote, off-grid areas in countries with relatively low electrification rates. A number projects under carbon finance are being planned in different ASEAN countries and feasibility studies are underway to investigate how much reduction in GHG could be made with the proposed projects [48, 101-103].

14. Conclusion

The results presented in the literature on the development of bioenegy in ASEAN and development of CDM projects in this part of the world reveal that this region of the globe could lead the world in bioenergy with a unified community where all member countries concentrate on collective resources of biomass; member countries (Malaysia, Indonesia, Thailand and The Philippines) share technological expertise with developing member countries (Cambodia, Lao PDR, Myanmar, Vietnam). Developed countries could provide training to cater a skilled workforce for the developing community and centralized research and development centres for biomass and bioenergy technologies. Singapore, and Malaysia could initiate in setting up bio-refineries and MSW treatment (waster-to-energy) plants; and regional collaboration on development and utilization of unified bioenergy resources. With these collective and integrated efforts this region would not only become energy sufficient using bioenergy resources but lead the world in this area. Lim and Lee [94] proposed a diamond framework for ASEAN biomass bioenergy cooperation that provides an ideal unified framework for this community to work together and this would lead the ASEAN countries towards leadership in bioenergy where the developing members as well as developed ones are to play their roles to achieve energy as well as social sustainability.

Acknowledgements

The author acknowledge the useful discussion with Dr. Lim Chee Ming and grateful to Dr. M. G. Blundell, Faculty of Science, University of Brunei Darussalam for providing valuable comments on the manuscript.

Author details

A. Q. Malik*

Address all correspondence to: Abdul.Malik@fnu.ac.fj, malikaqs@gmail.com

School of Pure Sciences, College of Engineering, Science & Technology, Fiji National University, Lautoka Campus, Fiji

References

[1] Carlos, R. M., D. B. Khang, Characterization of biomass energy projects in southeast Asia, Biomass and Bioenergy 32(2008) 525-532.

[2] Parnphumeesup, P., S. A. Kerr, Stakeholder preferences towards the sustainable development of CDM projects: Lessons from biomass (rice husk) CDM project in Thailand, Energy Policy (2011) 3591-3601.

[3] UNEP Report, Status and barriers of CDM projects in southeast Asian countries Available at: http://scholar.google.com.au/scholar?q=UNEP+Report,+Status+and+barriers+of+CDM+projects+in+southeast+Asian+countries&hl=en&as_sdt=0&as_vis=1&oi=scholart&sa=X&ei=iE8QUJ65McSPiAfvy4GgAQ&sqi=2&ved=0CE0QgQMwAA

[4] Ponlok, T., CDM development in Cambodia. Presented at fourth regional workshop and training on CD4CDM, 4-5 April 2005, AIT, Bangkok, Thailand.

[5] Bratasida, L., Regional CDM updates: selected ASEAN countries, Annex 1 Expert Group Seminar, OECD, 27-28 March 2006.

[6] Clean Development Mechanisms (CDM) in Lao PDR, Water resources and environmental administration (WREA), Department of environment, Prime Minister's Office of Lao PDR; 2011.

[7] Clean development mechanisms: Malaysia's experience, Conservation and environmental management division, Ministry of Science, Technology and the Environment, Malaysia. Viewed at: http://archive.unu.edu/update/downloads/RN_Report.pdf

[8] Peskett, L., J. Brown, K. Schreckenberg, Carbon offsets for forestry and bioenergy: Researching opportunities for poor rural communities, Final Report, May 2010.

[9] Chantanakome, W., Regional cooperation in promotion and sustaining CDM initiatives. Paper presented Asian regional workshop on "capacity development for the clean development mechanism (CD4CDM), 19-21 October 2005, AIT, Bangkok, Thailand.

[10] NEA. Clean Development Mechanism, National Environmental Agency (NEA), Government of Singapore Available at: http://www.google.com.sg/webhp?sourceid=navclient&ie=UTF-8

[11] Hoa, H. M., Potential CDM projects in Vietnam. Paper presented at workshop on the financing modalities of clean development mechanism (CDM) 27-28 June 2005, Jakarta, Indonesia.

[12] Malik, A. Q., Assessment of the potential of renewables for Brunei Darussalam, Renewable and Sustainable Energy Reviews 15(2011) 427-437.

[13] Ngoc., U.N., H. Schnitzer, Sustainable solutions for solid waste management in southeast asian countries, Waste Management 29(2009) 1982-1995.

[14] Top, N., N. Mizoue, S. Ito, S. Kai, Spatial analysis of woodfuel supply and demand in kampong thom providence, Cambodia, Forest Ecology and Management 194(2004) 369-378.

[15] Top, N., N. Mizoue, S. Ito, S. Kai, T. Nakao, S. Ty, Re-assessment of woodfuel supply and demand relationships in kampong thom province, Cambodia, Biomass and Bioenergy 30 (2006) 134-143.

[16] Top, N., N. Mizoue, S. Ito, S. Kai, T. Nakao, Variation in woodfuel consumption patterns in response to forest availability in kampong thom province, Cambodia, Biomass and Bioenergy 27 (2004) 57-68.

[17] Abe, H., A. Katayama, B. P. Sah, T. Toriu, S. Samy, P. Pheach, M. A. Adams, P. F. Grierson, potential for rural electricfication based on biomass gasification in Cambodia, Biomass and Bioenergy 31 (2007) 656-664.

[18] Koopmans, A., Biomass energy demand and supply for south and south-east asia – assessing the resource base, Biomass and Bioenergy 28 (2005) 133-150.

[19] Cambodia bioenergy development promotion project, Study Report, Prepared by Engineering and consulting firms association, Japan, Japan development institute (JDI) and Kimura chemical plants company limited; February 2007.

[20] Sukkasi, S., N. Chollacoop, W. Ellis, S. Grimley, S. Jai-In, Challenges and considerations for planning toward sustainable biodiesel development in developing countries: Lessons from the Greater Mekong Subregion, Renewable and Sustainable Energy Reviews 14 (2010) 3100-3107.

[21] Ministry of Agriculture, Forestry and Fisheries, The information centre on economics land concession in Cambodia, can be reviewed at: http://www.elc.maff.gov.kh/.

[22] Sasaki, N., A. Yoshimoto, Benefits of tropical forest management under the new climate change agreement – a case study in Cambodia, Environmental Science & Policy 13 (2010) 384-392.

[23] Suntana, A. S., K. A. Vogt, E. C. Turnblom, R. Upadhye, Bio-methonal potential in Indonesia: Forest biomass as a source of bio-energy that reduces carbon emission,

Applied Energy 86 (2009) 5215-5221. [A] Vogt, K. A., D. J. Vogt, T. Patel-Weynand, R. Upadhye, D. Edlund, R. L. Edmonds, Bio-methanol: how energy choices in the western United States can help mitigate global climate change. Renewable Energy 34 (2009) 233-241.

[24] FAO, Global forest resources assessment 2005: progress tpwards sustainable forest management, Rome, Italy: Forest department, Food and Agriculture Organization of United Nations (FAO); 2006.

[25] Suntana, A. S., E. C. Turnblom, K. A. Vogt, Addressing unknown variability in seemingly fixed national forest estimates: aboveground forest biomass for renewable energy, International Scientific Journal, submitted for publication

[26] Kumarwardhani, L., Global forest resources assessment 2005: Indonesia Country Report. Rome, Italy: Forestry Department, Food and Agricultural Organization of the United Nations (FAO); 2005.

[27] FAO. State of the world's forest 2005, Rome, Italy: Forestry Department, Food and Agriculture Organization of United Nations (FAO); 2005.

[28] Gunawan, S., S. Maulana, K. Anwar, T. Widjaja, Rice bran, a potential source of biodiesel production in Indonesia, Industrial Crops and Products 33 (2011) 624-628.

[29] Wicke, B., R. Sikkema, V. Dornburg, A. Faaij, Exploring land use changes and the role of palm oil production in Indonesia and Malaysia, Land Use Policy 28 (2011) 193-206.

[30] U.S. Department of Agriculture. Indonesia and Malaysia palm oil production; 2007. Available from: http://www.pecad.fas.usda.gov/highlights/2007/12/Indonesia_palmoil/

[31] Biopower Asia: Renewable Cogen Asia; 2010. Available at: http://www.rcogenasia.com/biomass-power-cogen-2/biopower-asia/

[32] Widodo, T. W., E. Rahmarestia, Current status of bioenergy development in Indonesia. Paper presented at regional forum on bioenergy sector development: challenges, opportunities, and the way forward, 23-25 January 2008, Bangkok, Thailand.

[33] ITA. Renewable energy market assessment report: Indonesia, U.S. Department of Commerce, International Trade Administration, Washington. Available: www.trade.gov.

[34] Jayed, M. H., H. H. Masjuki, M. A. Kalam, T. M. I. Mahlia, M. Husnawan, A. M. Liaquat, Prospectus of dedicated biodiesel engine vehicles in Malaysia and Indonesia, Renewable and Sustainable Energy Reviews 15 (2011) 220-235.

[35] Jupesta, J, Modeling technological changes in the biofuel production system in Indonesia, Applied Energy doi:10.1016/j.apenergy.2011.02.020

[36] Resturi, D., A. Michaelowa, The economic potential of bagasse cogeneration as CDM projects in Indonesia, Energy policy 25 (2007) 3952-3966.

[37] Das, A., E. O. Ahlgren, Implications of using clean technologies to power selected ASEAN countries, Energy Policy 38 (2010) 1851-1871.

[38] Sawathvong, S., Experiences from developing an integrated land-use planning approach for protected areas in the Lao PDR, Forest Policy and Economics 6 (2004) 553-566.

[39] FAO. Lao PDR forestry outlook study, APFSOS II/WP/2009/17, Bangkok: Food and Agricultural Organization Office for Asia and the Pacific (FAO); 2009.

[40] Decentralized biofuel supply chain development study in Lao PDR, Application of biofuel supply chains for rural development and Lao energy security measures, A joint study conducted by Engineering and consulting firms association, Japan, Japan development institute (JDI), Japan bioenergy development corporation (JBEDC) and Lao bioenergy corporation (LBEDC); 2008.

[41] Bush, S. R., Acceptance and suitability of renewable energy technologies in Lao PDR, Report for Asia Pro Eco Project TH/Asia Pro Eco/05 (101302); 2006.

[42] Dannenmann, B. M. E., C. Choocharoen, W. Spreer, M. Nagle, H. Leis, A. Neef, J. Mueller, The potential of bamboo as source of renewable energy in northern Laos. Paper presented at a conference on international agricultural research for development, University of Kassel-Witzenhausen and University of Gttingen, 9-11 October 2007.

[43] Wu, T. Y., Abdul Wahab Mohammad, Jamaliah Md. Jahim, Nurina Anuar, A holistic approach to managing palm oil mill effluent (POME): Biotechnological advances in the sustainable reuse of POME, Biotechnology Advances 27 (2009) 40-52.

[44] Yee, K. F., K. T. Tan, A. Z. Abdullah, K. T. Lee, Life cycle assessment of biodiesel: Revealing facts and benefits for sustainability, Applied Energy 86 (2009) 5189-5195.

[45] Shuit, S. H., K. T. Tan, K. T. Lee, A. H. K. Kamaruddine, Oil palm biomass as a suitable energy source: Malaysian case study, Energy 34 (2009) 1225-1235.

[46] Jayed, M. H., H. H. Masjuki, M. A. Kalam, T. M. I. Mahlia, M. Husnawan, A. M. Liaquat, Prospectus of dedicated biodiesel engine vehicles in Malaysia and Indonesia, Renewable and Sustainable Energy Reviews 15 (2011) 220-235.

[47] Goh, C. S., K. T. Lee, Palm-based biofuel refinery (PBR) to substitute petroleum refin ery: An energy and emergy assessment, Renewable and Sustainable Reviews 14 (2010) 2986-2995.

[48] IEA. Deploying renewables in Southeast Asia 2010, International Energy Agency (IEA); 2010.

[49] Chua, S. C., T. H. Oh, W. W. Goh, Feed-in tariff outlook in Malaysia, renewable and Sustainable Energy Reviews 15(2011) 705-712.

[50] Marchetti, J. M., A summary of the available technologies for biodiesel production based on a comparison of different feedstock's properties, Process Safety and Environmental Protection, doi: 10.1016/j.psep.2011.06.010.

[51] Goh, C. S., K. T. Lee, A visionary and conceptual macroalgae-based third-generation bioethanol (TGB) biorefinary in Sabah, Malaysia as an underlay for renewable and sustainable development, Renewable and Sustainable Energy Reviews 14 (2010) 842-848.

[52] Ahmad, A. L., N. H. Mat Yasin, C. J. C. Derek, J. K. Lim, Microalgae as a sustainable energy source for biodiesel production: A review, Renewable and Sustainable Energy Reviews 15 (2011) 584-593.

[53] Goh, C. S., K. T. Tan, K. T. Lee, S. Bhatia, Bio-ethanol from lignocelluloses: Status, perspectives and challenges in Malaysia, Bioresource Technology 101 (2010) 4834-4841.

[54] Ahmad, A. L., N. H. Mat Yasin, C. J. C. Derek, J. K. Lim, Microalgae as a sustainable energy source for biodiesel production: A review, Renewable and Sustainable Energy Reviews 15 (2011) 594-593.

[55] Balat, M., H. Balat, Progress in biodiesel processing, Applied Energy 87 (2010) 1815-1835.

[56] Singh, R. P., A. Embrandiri, M. H. Ibrahim, N. Esa, Management of biomass residues generated from palm oil mill: Vermicomposting a sustainable option, Resources, Conversation and Recycling 55 (2011) 423-434.

[57] Singh, R. P., P. Singh, A. S. F. Araujo, M. H. Ibrahim, O. Sulaiman, Management of urban solid waste: Vermicomposting a sustainable option, Resources, Conservation and Recycling 55 (2011) 719-729.

[58] Goh, C. S., K. T. Lee, Will biofuel projects in Southeast Asia become white elephants?, Energy Policy 38 (2010) 3847-3848.

[59] Gagandeep, K., The biogas kitchen. The Hindu business daily from the Hindu group of publications; 2007. Available at: http://www.blonnet.com/life/2007/01/26/stories/2007012600110200.htm.

[60] Tock, J. Y., C. L. Lai, K. T. Lee, K. T. Tan, S. Bhatia, Banana biomass as potential renewable energy resource: A Malaysian case study, Renewable and Sustainable Energy Reviews 14 (2010) 798-805.

[61] Swe, M. M., Entrepreneurship development in solar energy sector for rural area in Myanmar. Paper presented at ARTES/SESAM Alumini Level Workshop, 19-23 May 2008; Nepal. Available at: www.iim.uni-flensburg.de/sesam/upload/Asiana_Alumni/MyatMon.pdf

[62] Elauria, J. C., M. L. Y. Castro, M. M. Elaeria, S. C. Bhattacharya, P. Abdul Salam, Assessment of sustainable energy potential of non-plantation biomass resources in the Philippines, Biomass and Bioenergy 29 (2005) 191–198.

[63] Elauria, J. C., M. L. Y. Castro, D. A. Racelis, Sustainable biomass production for energy in the Philippines, Biomass and Bioenergy 25 (2003) 531-540.

[64] Manoscoc, M. L. C., A year after: Status of Philippines Biofuel Act, available at: www.unapcaem.org/publication/bioenergy.pdf

[65] Raymond, R. T., A. B. Culaba, M. R. I. Purvis, Carbon balance implications of coconut biofuel utilization in the Philippine automotive transport sector, Biomass and Bioenergy 26 (2004) 579-585.

[66] Mendoza, T. C., R. Samson, Relative bioenergy potentials of major agricultural crop residues in the Philippines, Philippine Journal of Crop Science 31 (2006) 11-28.

[67] Energy working group, Assessment of biomass resources from marginal lands in APEC economies, Asia-Pacific Economic Cooperation; 2009.

[68] Gadde, B., S. Bonnet, C. Menke, S. Garivait, Air pollutant emissions from rice straw open field burning in India, Thailand and the Philippines, Environmental Pollution 157 (2009) 1554-1558.

[69] http://en.wikipedia.org/wiki/Biodiesel

[70] APO. Solid-Waste management, Issues and challenges, Asian Productivity Organisation (APO), Tokyo; 2007.

[71] Wang, J,Y., Municipal organic waste as an alternative urban bioenergy source. Proceedings of the Regional Forum on Bioenergy Sector Development: Challenges, Opportunities and Way Forward, Bangkok, Thailand, pp.115-140; 2008. Available at: http://www.unapcaem.org/publication/pub_Bio.htm

[72] Khoo, H. H., T. Z. Lim, R. B. H. Tan, Food waste conversion options in Singapore: Environmental impacts based on an LCA perspective, Science for the Total Environment 408 (2010) 1367-1373.

[73] Integrated Solid Waste Management in Singapore, National Environmental Agency & Ministry of the Environment & Water Resources Singapore, ASIA 3R Conference, 30 Oct.-1Nov 2006.

[74] Sajjakilnukit B., R. Yingyuad, V. Maneekhao, V. Pongnarintasut, S. C. Bhattacharya, P. Abdul Salam, Assessment of sustainable energy potential of non-plantation biomass resources in Thailand, Biomass and Bioenergy 29 (2005) 214-224. [A] Yan, J., T. Lin, Biofuels in Asia, Applied Energy 86 (2009) S1-S10.

[75] Junginger, M., A. Faaij, R. Van den Broek, A. Koopmans, W. Hulscher, Fuel supply strategies for large-scale bio-energy projects in developing countries. Electricity generation from agricultural and forest residues in Northeastern Thailand, Biomass and Bioenergy 21 (2001) 259-275.

[76] Prasertsan, S., B. Sajjakulnukit, Biomass and bioenergy in Thailand: Potential, opportunity and barriers, Renewable Energy 31 (2006) 599-610.

[77] Krukanont, P., S. Prasertan, Geographical distribution of biomass and potential sites of rubber wood fired power plants in Southern Thailand, Biomass and Bioenergy 26 (2004) 47-59.

[78] Silalertruksa, T., S. H. Gheewala, M. Sagisaka, Impacts of Thai bio-ethanol policy target on land use and greenhouse emissions, Applied Energy 86 (2009) 5170-5177.

[79] DEDE, www.dede.go.th Available at: www.dede.go.th/dede/fileadmin/usr/bers/gasohol_2008/3-5111113_Monthly_selling_gasohol_47_51.xls

[80] Silalertruksa, T., S. H. Gheewala, Enviromental sustainability assessment of bio-ethanol production in Thailand, Energy 34 (2009) 1933-1946.

[81] Permchart, W., V. I. Kouprianov, Emission performance and combustion efficiency of a conical fluidized-bed combustor firing various biomass fuels, Biosource Technology 92 (2004) 83-91.

[82] Janvijitsakul, K., V. I. Kurprianov, Major gaseous and PAH emissions from a fluidized-bed combustor firing rice husk with high combustion efficiency, Fuel Processing Technology 89 (2009) 777-787.

[83] Shrestha, R. M., S. Malla, M. H. Liyanage, Scenario-based analyses of energy system development and its environmental implications in Thailand, Energy Policy 35 (2007) 3179-3193.

[84] Delivand, M. K., M. Barz, S. H. Gheewala, B. Sajjakulnukit, Economic feasibility assessment of rice straw utilization for electricity generating through combustion in Thailand, Applied Energy 88 (2011) 3651-3658.

[85] Sawangphol, N., C. Pharino, Status and outlook for Thailand's low carbon electricity development, Renewable and Sustainable Energy Reviews 15 (2011) 564-573.

[86] APEC, Assessment of biomass resources from marginal lands in economies, Asia-pacific Economic Cooperation (APEC) 2009. [A] Hanh, V. T, L. T. B. Phuong, L. T. Hung, Sustainable ethanol production in Vietnam: Current status and proposed solutions (2012) available at: http://www.ibt.vn/en/32 sustainable ethanol production in Vietnam: current status and proposed solutions. [B] Oguma, M., Y. J. Lee, S. Goto, An overview of biodiesel in asian countries and the harmonization of quality standards, International Journal of Automotive Technology 13(1) (2012) 33-41.

[87] Dinh Man, T. Biofuel production from biomass and the state bio-energy development program of Vietnam, presented at: Biomass-Asia Workshop, Institute of Biotechnology Vietnamese Academy of Science and Technology Guangzhou, China 2008.

[88] Nguyen, N. T., Rice husk potential of Vietnam, presented at Biomass-Asia Workshop, Bangkok, Thailand, December 13-15, 2005.

[89] Tuyen, T. M., Biomass utilization in Vietnam, presented at Biomass-Asia Workshop, Bangkok, Thailand, December 13-15, 2005.

[90] Dang, T. T., O. Saito, Y. Yamamoto, A. Tokai, Scenarios for sustainable biomass use in the Mekong Delta, Vietnam, journal of Sustainable Energy and Environment 1 (2010) 137-148.

[91] Nguyen, N. T., M. Ha-Duong, Economic potential of renewable energy in Vietnam's power sector, Energy Policy 37 (2008) 1601-1613.

[92] Shi, A. Z., L. P. Koh, H. T. W. Tan, The biofuel potential of municipal solid waste, GCB Bioenergy (2009), doi:10.1111/j.1757-1707-2009.-1024.x.

[93] Kofoworola, O. F., S. H. Gheewala, Estimation of construction waste generation and management in Thailand, Waste management 29 (2009) 731-738.

[94] Lim, S., K. T. Lee, Leading global energy and environmental transformation: Unified ASEAN biomass-based bio-energy system incorporating the clean development mechanism, Biomass and Bioenergy (2011) doi:10.1016/j.biombioe.2011.04.013.

[95] Sasaki, N., W. Knott, D. R. Foster, H. Etoh, H. Ninomiya, S. Chay, S. Kim, S. Sun, Woody biomass and bioenergy potentials in Southeast Asia between 1990 and 2020, Applied Energy 86 (2009) 5140-5150.

[96] Mustonen, S. M., Rural energy survey and scenario analysis of village energy consumption: A case study in Lao People's Democratic Republic, Energy Policy 38 (2010) 1040-1048.

[97] IEA. Oil Market Report December 2009, OECD/IEA, Paris; 2009.

[98] Arrieta, F. R. P., F. N. Teiceira, E. Yáñez, E. Lora, E. Castillo, Cogeneration potential in the Columbian palm oil industry: Three case studies, Biomass and Bioenergy 31 (2007) 503-511.

[99] Charles Edric, T. Co, Tan, M. C., Diamante, J. A. R., Yan, L. R. C., Tan, R. R., Razon, L. F., Internal mass-transfer limitations on the transesterification of coconut oil using anionic ion exchange resin in a packed bed reactor, Catalysis Todey (2011) doi: 10.1016/j.cattod.2011.02.065.

[100] Lin, L., Z. Cunshan, S. Vittayapadung, S. Xiangqian, D. Mingdong, Oppertunities and challenges for biodiesel fuel, Applied Energy 88 (2011) 1020-1031.

[101] UNFCCC, United Nations Framework Convention on Climate Change (UNFCCC) Project Activities Search Database; 2009, available online: http://cdm.unfccc.int/Projects/projsearch.html.

[102] IEA, Energy Balances of Non-OECD Countries (2009 Edition), IEA/OECD, Paris; 2009.

[103] Parnphumeesup, P., S. A. Kerr, Classifying credit buyers according to their attitudes towards and involvement in CDM sustainability levels, Energy Policy (2011) doi: 10.1016/j.enpol.2011.07.026.

Chapter 3

Development of the Technology for Combustion of Large Bales Using Local Biomass

Branislav S. Repić, Dragoljub V. Dakić,
Aleksandar M. Erić, Dejan M. Đurović,
Stevan D. J. Nemoda and Milica R. Mladenović

Additional information is available at the end of the chapter

http://dx.doi.org/10.5772/51095

1. Introduction

In terms of sustainable energy development in Serbia, as well as in the whole world, there is a growing need for using the alternative energy sources. Alternative energy sources are, in most cases, renewable: biomass, wind power, solar energy, hydro-power and geothermal energy. A need for the utilization of this kind of energy sources is dictated by the market, on one side, as well as by environmental protection, on the other. Prices of fossil fuels grow proportionally to the decreasing of fossil fuel reserves. Since available reserves of fossil fuels in Serbia, especially those of high quality, are relatively limited, this problem becomes even more emphasized [1-3]. On the other hand, it is necessary to harmonize the energy production legislation and practice in Serbia with the directives of the European Union, in the sense of intensifying the utilization of renewable energy sources and thus reducing pollution and greenhouse effect formation.

Biomass is one of key renewable energy sources [4]. This is the reason for the development of cheap thermal devices (boilers and furnaces) burning biomass from agricultural production as quite available and cheap energy source. These devices could be used primarily in villages, small towns and small businesses processing agricultural goods (greenhouses, dairy farms, slaughterhouses etc.) [5]. The devices could also be used for heating schools, hospitals, prisons and other institutions.

Annual energy consumption in the Republic of Serbia currently reaches 15 million tons oil equivalent (Mtoe), out of which 7.4 Mtoe represents the net consumption and 3 Mtoe is electricity consumption. According to the official date of Ministry of Infrastructure and Energy of Republic of Serbia [6], Serbia is in dispose of 4.3 Mtoe of renewable energy sources, while biomass is represented with 2.7 Mtoe. 60% out of registered biomass potential are residuals from agricultural production, and the rest is wood biomass. Currently, only a small portion of waste biomass is being used in energy production mostly for heating (not taking into account burning in the individual households, in small ovens), for several reasons: low electricity price and non-resolved problems in biomass gathering. Also, there is no regulated biomass market, and no developed technologies for its utilization as fuel. Besides, small financial power of potential buyers have to be mentioned, as well as costly commercial credits and total absence of state subsidizing of biomass facilities.

This biomass is a cheap and available fuel, but its utilization is linked to the problems of its collection, preparations for its transportation (cutting, tying into haystacks, baling), transportation and storage [7]. The best way for utilizing residual agricultural biomass for energy production in industrial or district heating is to be used close to place of its gathering - in large agricultural companies. That is the optimal solution, from energy, as well as economic point of view. Agricultural biomass is usually collected in form of bales, varying in size and shape, so it is most convenient to use it in that form. One of the most efficient ways, recommended by many institutions worldwide, is the combined heat and power (electricity) production – CHP [8], which use residual biomass as fuel, and have least as possible own power consumption.

Two technologies are currently used for the combustion of biomass bales. The first is based on whole-bale combustion in the combustion chamber, while the second considers combustion of biomass bales in "cigar" burners. The "cigar" firing technology provides better quality of the combustion process, resulting in lower pollutant emissions and increased plant efficiency. This technology was found to be very suitable for straw combustion and was deemed not to be associated with any process limitations.

2. Research path

In the process of production of different crop residues that occur in some species exceed three times the amount of crops produced. These residues can be baled and still used. Today in Serbia the producers of baled biomass are mainly farmers who receive biomass in baled form as by-products of primary production. In practice there are two basic types of bales: small square bales (0,40x0,50x0,80 m) and large cylindrical bales (ø1,80x1,20 m) or rectangular shape (0,80-1,20x0,70x1,50-2,50 m).

Advantages and disadvantages of specific types of bales are the following [9]. Small conventional bales have many advantages: low cost of presses, binders moderate prices, the need for a smaller tractor, good storage, a favorable agreement on means of transport, simple disintegration and chopping by means of lower prices, the possibility of firing the entire bales and others. Deficiencies are inevitable manual operation, by hand using auxiliary storage means, a relatively high usage of the binder, the lower reliability than other presses etc.

The advantages of large cylindrical bales are moderate presses price, simple and fully mechanized manipulation, in the case of unwinding a simple and inexpensive device, conveniently storing for own needs on medium farms, the opportunity to work with medium power tractors etc. The disadvantages of this bundle are: the highest consumption of binder, the lower performance because of the need to halt the bale tying and ejection from the workspace, the sensitivity of trusses on the quality of the binder, the deformation under the bonding quality, lower transportability because the empty space, need more storage space etc.

Large square bales have the following advantages: high pressure compression, high performance, low consumption of binder, best transportability, good storage conditions, the whole mechanization and the lowest price of manipulation, the lowest consumption of binder etc. The disadvantages are reflected in the following: high initial cost of machinery, required a large tractor, requires special means for manipulation, machinery sensitive to the application of low-quality binders, need of special funds for the disintegration etc.

Furnaces and boilers that would use baled biomass from agricultural production can be a wide power range from 0.1 to 2 MW or more. Baled biomass as a fuel does not require big investments in preparation because balers have nearly every farmer. These are not expensive and complicated machines and do not require high energy consumption per kg of baled biomass. On the other hand, neither of which would be the biomass used as fuel, not far from the place of origin nor transport is not a major problem. Storage can be a problem, but as there is plenty of farmland damage caused by his occupation is insignificant, and increased investment costs to build a warehouse to quickly pay the difference in price between the liquid or gaseous fuels and biomass. In addition, profit get from the green credits - benefits that are obtained in the case of renewable energy usage are paid this facilities only through the benefits in a very short period of time.

In the Laboratory for Thermal Engineering and Energy of the "Vinca" Institute in Belgrade, efforts have been made to develop a clean technology for utilizing baled biomass for energy production. The initial set of analyses carried out in the research investigation conducted focused on the combustion of small, 40x50x80 cm straw bales in cigar burners. For the said purpose, an experimental, 75 kWth hot water boiler was designed and constructed [9]. The furnace was built entirely out of an insulating material providing favorable biomass combustion conditions. Appropriate boiler tests were conducted in order to properly determine required design parameters. Although the boiler assembly examined was a small-scale facility intended to be used by individual farm owners and utilized for space heating, it provided a good basis for development of large, industrial scale straw-fired facilities.

Following the initial set of analyses, combustion of rolled, ø180x120 cm straw bales in cigar burners was analyzed in the next investigation phase. In order to assess the combustion quality and obtain data needed for proper design of the straw-fired hot water boiler, a 1 MWth demonstration furnace was designed, constructed and tested [10].

As a result of the specified investigation efforts, a pilot plant capable of burning large, 0.7x1.2x2.0 m straw bales was designed and built [11]. A 1.5 MWth industrial-scale hot water boiler was constructed and installed in the Agricultural Corporation Belgrade, where it was used for heating 1 ha of vegetable greenhouses belonging to the agricultural complex mentioned. The boiler house was built in the immediate vicinity of the greenhouse complex.

3. Materials and methods

Technologies enabling biomass use for energy generation are mainly dependant on biomass characteristics. Different biomass conversion technologies available on the market include: fixed-bed combustion, combustion on the grate, combustion in dust burners, fluidized bed combustion and gasification [12]. Utilization of agricultural biomass faces a lot of challenges. One of the main disadvantages associated with combustion of agricultural biomass is a tendency of biomass ash to melt [13]. Two technologies are currently used for the combustion of biomass bales. The first is based on whole-bale combustion in the combustion chamber, while the second considers combustion of biomass bales in so called "cigar" burners. Cigar burner technology was found to be very suitable for straw combustion and was deemed not to be associated with any process limitations. The research investigation described herein was focused on developing a cigar burner combustion system suitable for the combustion of bales of various sizes and shapes and their utilization for energy production.

3.1. An efficient boiler burning small straw bales

The experimental boiler burning small soya, corn, rape seed or wheat straw bales, with 0.8x0.5x0.4 m in size, has been designed and built [10, 14-16]. The combustion has been organized on the principles of cigarette burning [17]. Thermal power of the boiler is around 75 kW. In Figure 1, the scheme of the experimental boiler is shown. Baled straw is introduced through the inlet (3) into the combustion zone (7). The inlet is supplied by devices for continuous bale feedings and provides stable combustion conditions (Figures 2 and 3). Furnace walls (4) have been made of refractory material – chamotte, with thermal insulation (5).

Under the original solution fresh air is injected through two channels, the primary air through channel (8), and secondary air through channel (9), and they are divided using compartment (11). The tertiary air is supplied through the inlet (12), and is previously heated by flowing inside the walls (13). In the zone (14) is carried out the process of final combustion of the bale.

After the first examination of the boiler some changes were carried out in the distribution of air so that after this change the air for combustion is inserted into the space through the distributor (26) which is connected to a fan of fresh air (27). By changing the position of the air distributor can be regulated the part of the bale involved in combustion and thus indirectly is regulated the heat output of the boiler.

The heat produced by combustion of biomass is transferred by the gas-to-water heat exchanger (15). After passing through the channels (16) to the flue gases collector (17), the flue gases leave the boiler through the smokestack (18), equipped with the valve (19) and flue gas fan (28), and through the cyclone-type particle precipitator (29). Ash is collected in ash collectors (20, 21, and 22). A mobile tube for ash removal (23) has been placed inside the furnace, as well as a tube for pneumatic transport of ash (24). The boiler has a revision opening (25) for manual ash removal.

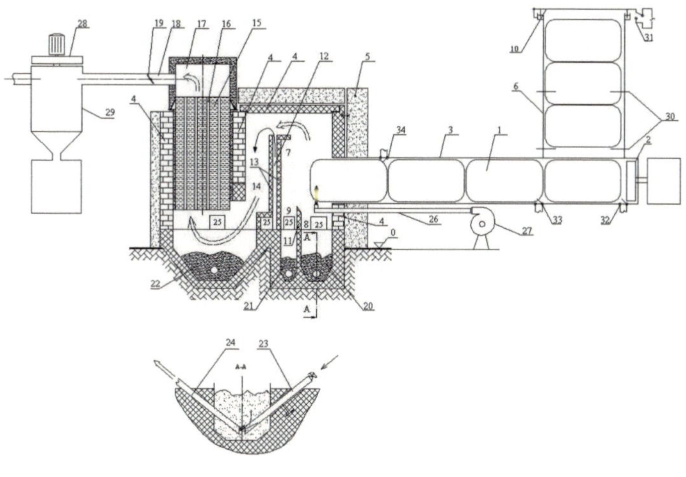

Figure 1. Scheme of the small agricultural biomass bale combustion boiler

Figure 2. Heat accumulator and bale storage and feeding system

Figure 3. Bale storage and feeding system with push piston

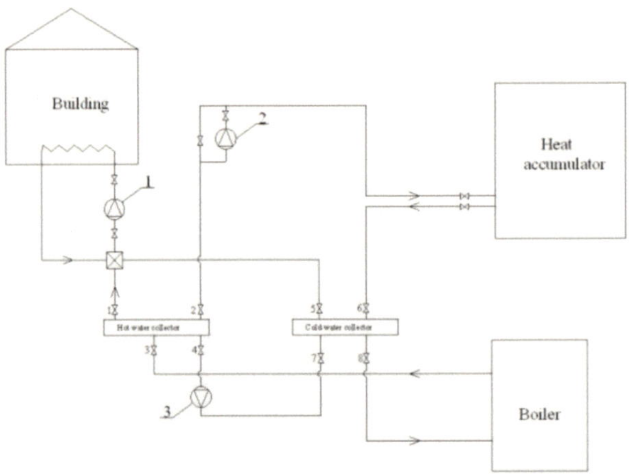

Figure 4. Thermal scheme of distribution facilities

In order to obtain to plant work at nominal power, heat accumulator (thermal reservoir, with volume of 5 m^3) has been installed (Figure 2). In this way it is ensured that no matter what the current needs for heating buildings are, boiler always works with the nominal power. The transitional periods (spring, autumn), for example, the need for heating usually amount to 20-40% rated power boiler, which would mean a much lower level of utility plant. Thermal scheme of distribution facilities is shown in Figure 4. From it can be seen following thermal circles: a) Hot water from the boiler goes directly into a building that is heated, b) Hot water from the boiler goes into heat only tank, c) Hot water from the boiler going at the same time in the building and heat reservoir, d) Hot water tank from the heat goes into the building. Also, the boiler is equipped with appropriate management and control system (Figure 5).

The thermal power of the boiler has been regulated with: the amount of straw engaged in the combustion process, the air excess and the fuel feeding rate. This experimental boiler could be scaled, since it satisfies the similarity requirements in: geometry, flow patterns, thermal load, thermal flux, adiabatic temperature, average temperature and flue gases content.

3.2. The demo furnace burning soya straw bales

In order to assess the combustion quality and to obtain data for the design of a soya straw-fired hot water boiler, a demo furnace with thermal power of 1 MW has been designed and built [10, 18, 19]. The appearance of the furnace, with the thermocouple probes, the primary air fan and channel, and the fuel feeding channel is shown in Figure 6. This furnace has been adopted for cylindrical bales, with 1.2-1.5 m in diameter which were available at that time. The cross is clearly visible on Figure 7 where the scheme of the experimental demonstration unit for burning large rolled soy straw bales was presented. There can be also clearly distinguish three characteristics combustion zones in the cigar burner: drying zone (6), zone of devolatilization (5) and zone of char burning (13).

Figure 5. The boiler control system and cyclone-type particle precipitator

The proximate analysis of soya straw used in testing is given in Table 1. The sum of five tests was done. A summary of main test parameters is given in Table 2. During all tests, three gas temperatures in the combustion zone were measured, with shielded type K thermocouple probes. Gas sampling was done with a probe placed near the furnace exit. Gas samples were continuously analyzed with two analyzers, collected every 5 seconds and stored on-line.

It should be noted that secondary air supply through the movable cross was not present in the first version of the demo furnace, which was examined in tests 1 and 2. The results from these tests stressed the need to introduce secondary air in the combustion zone, at the bale forehead, and the furnace with secondary air supply through the cross was examined in tests 3, 4 and 5.

Figure 6. The appearance of the demo facility for burning large rolled straw bales

Test 1 was conducted with one bale of straw placed in the feeding channel. Only temperature measurements were done, and the results showed that the temperature in the combustion zone, in steady conditions, was quite stable (730-830°C, Figure 8) for a reasonable period of time (40 minutes). It was noted that the amount of tertiary air did not contribute much to overall combustion conditions, and that in fact this air over-cooled the flue gases in the combustion zone.

Moisture (%)	Ash (%)	Char (%)	Fixed carbon (%)	Volatile matter (%)	Combustible matter (%)	Net calorific value (kJ/kg)
18.80	5.66	22.12	16.46	59.08	75.54	13686

Table 1. The proximate analysis of soya straw used in the tests

In test 2, along with temperatures, gas composition was continuously measured. Less air was supplied as tertiary than in test 1. In the initial, start-up period (Figure 9), gas samples were taken directly from the combustion zone, and very high levels of CO in the flue gases were noted. After the choking of the gas sampling probe and its cleaning, and also in all following tests, gas samples were taken only from the top of the furnace. As the temperature in this period increased to approximately 1000°C, bale feeding was slowed down, and this corresponds to the temperature downfall (min. 50-75). Soon after that, stable conditions were obtained (Figures 9 and 10), primarily by adjusting bale feeding.

Test	1	2	3	4	5
Number of bales in the feeding tube	1	2	2	3	3
Amount of straw [kg]	134,6	280	327,97	458,3	554,9
Primary air [m³/h]	1548	1548	1548	1350	1404**
Secondary air [m³/h]	-	-	234	418,25	228,14**
Tertiary air [m³/h]	504	252	108	259,2	259,2**
Calculated thermal power [kW]⁺	485,2	529,3	556,5	551,7	455,7
Average air excess coefficient λ [-]	not measured	4,71	2,61	4,12*	2,92**
Test duration [min]	47	89	99	140	205

Conditions: ⁺ - The thermal power was calculated over the entire test period
* - The air excess coefficient in Test 4 was calculated for the period shown in the diagrams (Figures 13 and 14)
** - The air flow rates refer only to the period shown in the diagrams (Figures 15 and 16), since in test 5 variable speed drives were used for changing the speed of the fans. The air excess coefficient λ was calculated for the same period

Table 2. Test parameters

High level of CO concentration at the furnace top in test 2 urged the introduction of a small amount (approximately 10% of total air) of secondary air in the combustion zone, which would cool down the movable cross at the same time. It was also noted that tertiary air flow rate should be decreased, and therefore secondary air was introduced to the detriment of tertiary air. This change in design was examined in test 3, with two bales placed in the feeding channel.

The supply of the secondary air through the cross provided excellent conditions for combustion (Figure 11) – the concentration of CO was equal to zero for most of the time during the test. The air distribution (82% primary air, 12% secondary, 6% tertiary) was found to be well suited for maintaining steady conditions inside the furnace. On the other hand, the stability of the thermal output was found to depend largely on the active length of the bale immersed into the furnace.

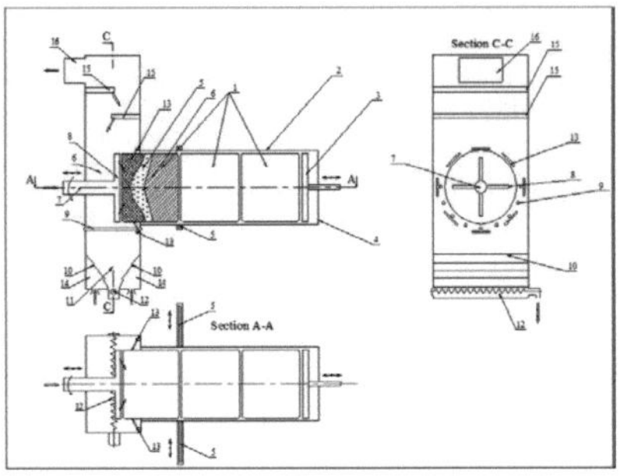

Figure 7. Schematic of the experimental demonstration unit for burning large rolled straw bales

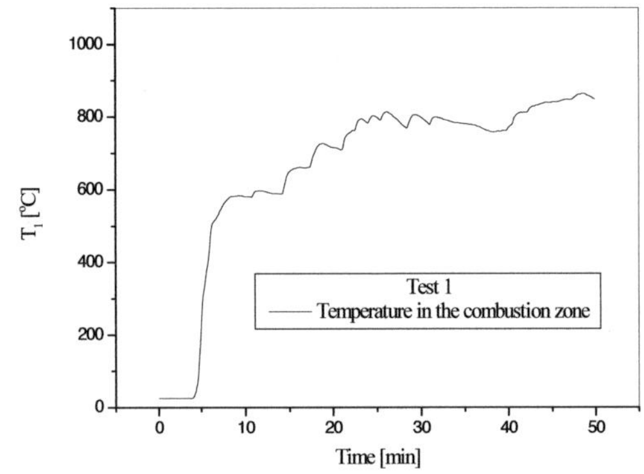

Figure 8. Test 1 – Temperature in the combustion zone

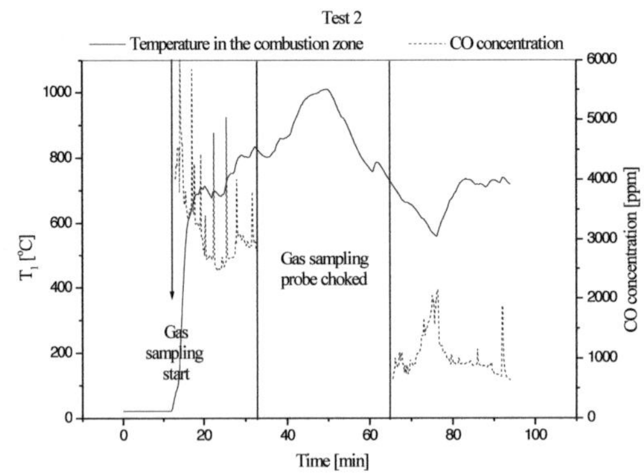

Figure 9. Test 2 – Temperature in the combustion zone vs. CO concentration

http://dx.doi.org/10.5772/51095

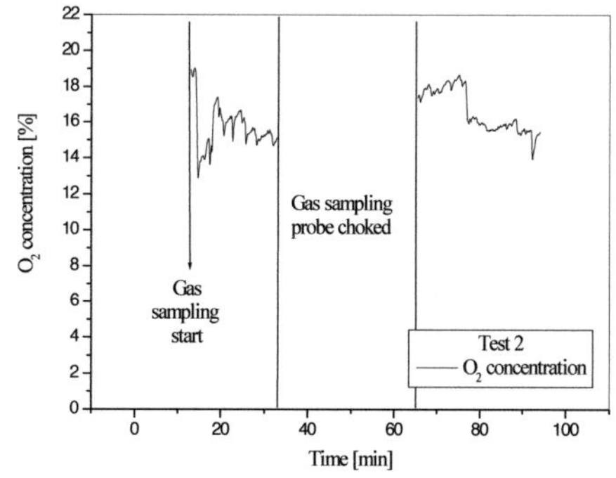

Figure 10. Test 2 – O_2 concentration at the furnace exit

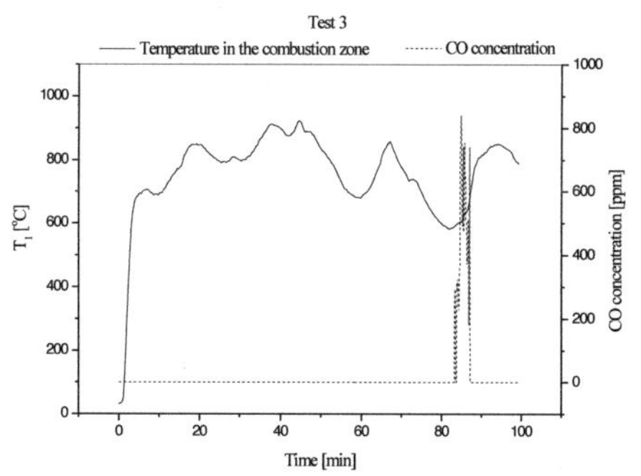

Figure 11. Test 3 – Temperature in the combustion zone vs. CO concentration

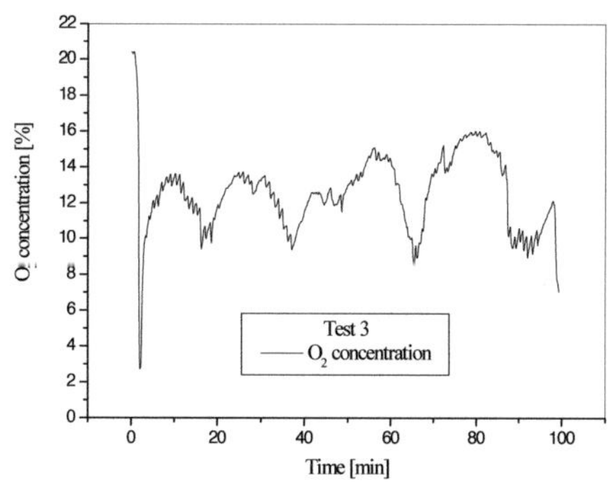

Figure 12. Test 3 – O_2 concentration at the furnace exit

Therefore, it is of great importance to feed the bale uniformly in accordance with the combustion process, and to maintain this length as stable as possible, by moving the cross accordingly. The temperature instabilities (from the minute 45 further on, Figure 11) during this test are a consequence of changes of this length. The only peak in CO concentration coincided expectedly with low temperatures during this period. Nevertheless, this test proved that the adopted concept of the furnace provided good conditions for efficient combustion of soya straw bales, with O_2 concentration ranging from 10-14% (Figure 12), and an optimal average value of λ.

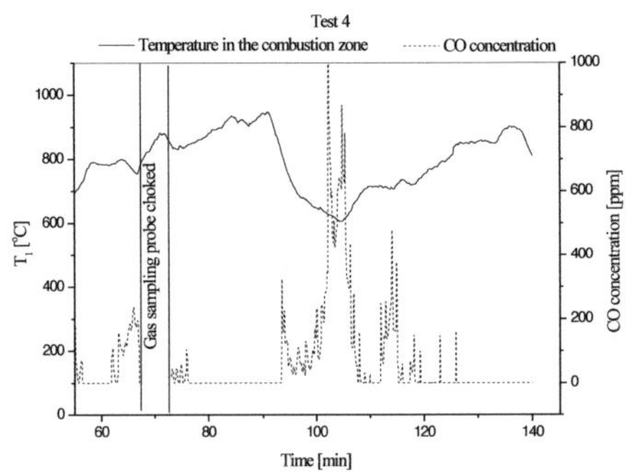

Figure 13. Test 4 – Temperature in the combustion zone vs. CO concentration

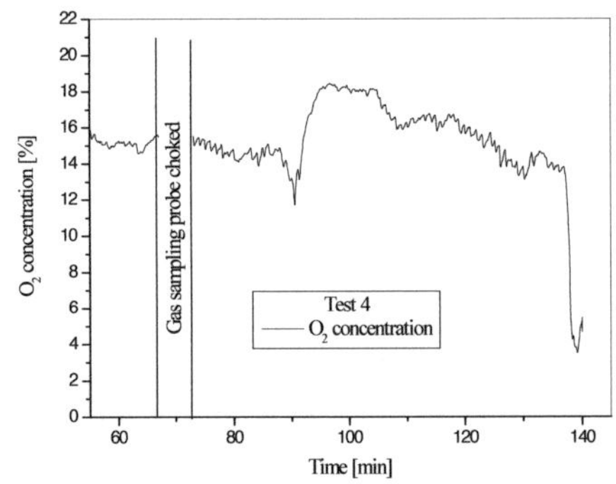

Figure 14. Test 4 – O_2 concentration at the furnace exit

The principal aim of test 4 was to assess the possibility of longer furnace operation, with three bales placed inside the feeding channel. The bales prepared for this test were approximately 1.2 m in diameter, and in order to secure stable manual feeding, the gaps between the channel wall and the bales were manually filled with more straw. Problems with feeding undersized bales caused instabilities in the first hour of the test. In the period shown in Figures 13 and 14, the temperature was in the desired range, and CO concentration was acceptable for most of the period (up to 350 ppm), the only rise in CO occurring at the time of the temperature downfall (minutes 100-110). It was spotted by visual inspection, through the inspection openings, that the bale was not inside the furnace at the time of the downfall, due to the problems with manual bale feeding and cross positioning – the bale forehead remained inside the tube. This caused the flame to enter the tube at the time, which also occurred during test 5.

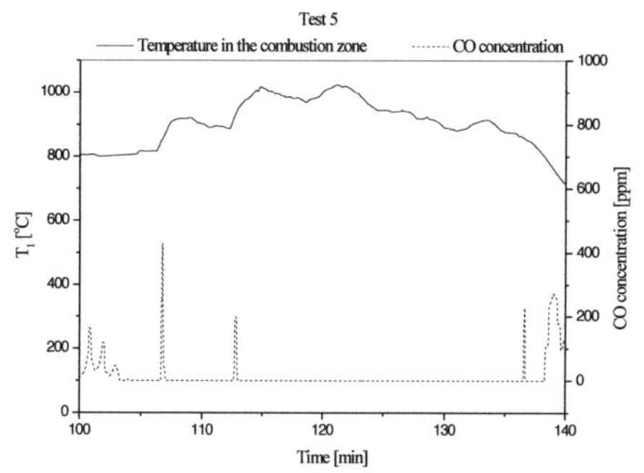

Figure 15. Test 5 – Temperature in the combustion zone vs. CO concentration

The aim of test 5 was to assess the influence of air flow rate control, with variable speed drives, on furnace performance. During a chosen period of 40 minutes (Figures 15 and 16), optimal air flow rates were obtained and bale feeding was kept stable. The concentration of CO was very low, with O_2 concentration varying in the range of 10-15%. The temperature during this period was higher than the desired 850°C (which should not be exceeded in order to avoid ash melting), which will be taken into consideration in some of the conclusions.

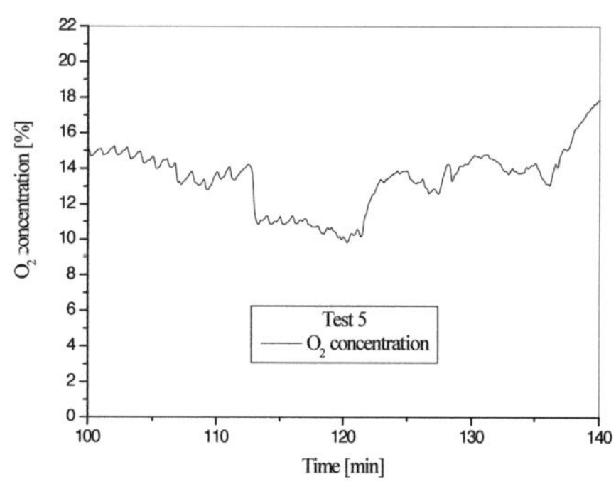

Figure 16. Test 5 – O_2 concentration at the furnace exit

3.3. Heat water boiler burning baled biomass

The use of renewable energy sources is becoming more and more important, mainly due to continuously increasing prices of fossil fuels, resource depletion and global attempts to achieve maximum feasible CO_2 emission reduction. Researches in this area are very complex and in order to obtain reliable data it is necessary to carry out theoretical and experimental research of the process. For this purpose, a 1.5 MW industrial-scale hot water boiler was constructed and installed in the Agricultural Corporation Belgrade [20-22]. The boiler is based on waste baled soybean (and other types) of straw combustion, and it is used for heating 1 ha (10000 m^2) of greenhouses. Combustion in the boiler carried on so-called "cigarette" principle [11], where 0.7x1.2x2.0 m straw bales are used as fuel. The bales have parallelepiped shapes. In Figure 17, the scheme of the experimental hot water boiler is shown.

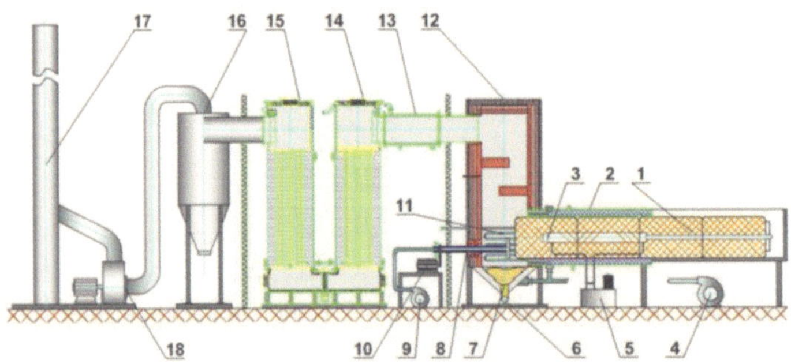

Figure 17. The scheme of the demonstrating hot water boiler with thermal power of 1.5 MW

Baled straw (Figure 17, position 1) is fed to the facility by cylinder type transporters (2,3). After entering the rectangle cross sectioned bale feeding channel (4) bales are carried by a motor driven VSD controlled conveyor (6) towards the furnace (7). The section of the channel (4) nearest to the furnace is made of multiple steel sheets (5), with primary air flowing through space between the sheets, thus cooling the sheets and being preheated at the same time. The furnace is made of refractory material (chamotte) (8), and is completely insulated (9). Ash is removed from the furnace by a transporter (10).

Preheated primary air (13) is supplied around the bale, and a portion of it also from under the grate (12), which is water cooled. Secondary air (11) is supplied through the movable cross onto the bale forehead, similarly as in the experimental furnace. The cross also serves as bale support from the forehead, for shaking-off ash from the forehead and for bale positioning inside the furnace.

Leaving the furnace chamber, the flue gases pass through a heat pipe to the first section of the gas to water heat exchanger (14), and then through a chamber with screen barrier type particle separator (15) to the second section of heat exchanger (16). After final particle removal in the multi-stage cyclone type separator (18) the flue gases, transported by the flue gas fan (19), leave the furnace through the stack (20). The view of the boiler house and heat accumulator is presented in Figure 18.

Apart from the presence of the movable cross, used for secondary air supply, which can be considered as innovative, another new concept is the existence of a 100 m^3 heat storage vessel - heat accumulator (Figure 18) with thermal insulation. It was introduced so that the whole facility could respond more appropriately to the heating needs of the greenhouses. Hot water produced in the boiler is stored in the heat storage vessel. At time when the ambient temperature is relatively high and weather conditions are mild (sunny days, without wind etc.), the boiler produces much more heat i.e. hot water than necessary for greenhouse heating. The greenhouse systems use only the amount of hat water necessary for heating, and the heat surplus is stored inside the heat storage vessel. At time when outside temperatures are below zero, on windy and cloudy days, the heat produced by the boiler might not be sufficient, but the lacking heat is then supplied from the heat storage vessel. The boiler is operated and controlled by a SCADA-based system, through a computer.

Figure 18. The boiler house, heat accumulator and cyclone type separator

A cigar firing combustion system is expected to exhibit the following advantageous features: a) combustion of whole bales and whole energy crops; b) compact combustor design; c) short start up period, good load-following performance; d) profitable operation of smaller facilities (down to 1 MWth); e) division of combustion from the heat recovery system, usable not only for the provision of steam (for heat generation or CHP), but also as a hot gas generator in industrial drying applications. Cigar burner combustion system promises a more competitive use of renewable for "green" heat and power generation as well as their use in various industrial applications.

Possible disadvantages of cigar burner combustion system include: a) a need for a "smart" and sophisticated process control system; b) thermal cracks, thermal attacks on the metal combustion chamber.

4. Experimental research

4.1. Investigation of the influence of bale quality

During experimental investigation of the boiler occasionally came to some minor problems in boiler operation. The problem was detected in the poor biomass burning. It is assumed that the main cause of problems is uneven quality of bales. Therefore, it is examined in detail the quality and moisture in bales that are stored and used in regular plant operation. It is assumed that poor bales quality could come from two reasons: a) Because of the rainy season in the period of collection of soybean straw in the fields; b) The increase in moisture content during bales storage up to their use.

Figure 19. The cover and open bales storage near the boiler house

Preparation for use baled straw in boiler is in September. According to official data of the Republic Hydrometeorological Service of Serbia [23] during the month of September at the territory of Belgrade fell more than 3 l/m^2 of rain. On this basis it can be concluded that the formed bale of soybean straw were acceptable dry before storing, or it can be said that the weather was ideal for baling straw. Shortly after baling bales were transported near the boiler room so that time needed to bales transport did not affect the increase in moisture.

5	10	15	20
4	9	14	19
3	8	13	18
2	7	12	17
1	6	11	16

Figure 20. Schematic layout of sampling straw

Part of the bales is stored under the canopy capacity of 1200-1500 pcs bundle (Figure 19). The quantity of bales is insufficient to operate the boiler throughout the season. Therefore it had to accede to the formation of a group bales in the open, again near the boiler where the bales are placed in the open and covered with nylon.

Number of sample	% of moisture	Number of sample	% of moisture
1	14.77	12	13.78
2	45.17	13	19.44
3	11.83	14	12.34
4	15.84	15	59.73
5a	66.70	16	17.59
5b	65.83	17	16.98
6	17.17	18	20.65
7	19.63	19	15.75
8	20.05	20a	68.98
9	16.74	20b	66.62
10	62.15	21	67.86
11	11.97		

Table 3. Results of the determination of moisture content in samples of soybean straw

From several groups of bales placed at open space one was chosen for the implementation of a test. Selected group made a bundle so that in a horizontal row was passed four bales (bale lying on the site with the largest surface area). Such orders were five in height. Length of group was determined free space and it corresponded to the width of a few dozen bales. It was decided to analyze the quality of bales (determination of moisture content in bales) per cross-section of group. A special sampler which allows sampling in-depth of the bale was made. Straw sampling scheme in the cross section is given in Figure 20 where from some of the bales taken more than one sample.

Determination of moisture content in straw samples was performed in an accredited laboratory for testing fuels at the Vinca Institute. Test results of straw samples moisture is given in Table 3. The results show dramatic differences in the quality of the bales at the cross section of bale group. All straw samples from the fifth highest among the group (number of samples 5a, 5b, 10, 15, 20a, 20b, 21) show that the concentration of moisture in them is extremely high. It ranged from 60-70%. Such quality straw with so much moisture content, absolutely can not burn in any furnace. Also, if such a bale enters the combustion chamber can cause a host of other problems.

By using the appropriate computer program calculation the adiabatic combustion temperature and combustion product composition of soybean straw with different moisture content

(Table 4). Results of proximate analyze of soybean straw was used as input data for calculations. In the case of combustion of soybean straw, with a total moisture content of Wt = 68%, the theoretical calculation shows that the combustion temperature is not possible to achieve satisfactory gas temperature for a real excess air to be used in the process of bales burning. The theoretical combustion temperature of soybean straw for excess air of 2.80 would be 572°C. In the case of combustion of the same composition biomass, but reduced the moisture content of Wt = 15%, the calculation shows that it is possible to achieve significantly higher gas temperature for a real excess air. In this case, the theoretical combustion temperature was 903°C for the excess air of 2.80. Note that this is mathematical calculated theoretical combustion temperature of soybean straw, which in real terms of furnace can not be achieved so that the actual combustion temperature significantly lower. This is caused by the fact that the combustion of CO to CO_2 achieves at the minimum temperature of 680°C. When burning soya straw with high moisture content in the flue gases products is a large amount of CO which is due to low temperatures (below 680°C) can not be transformed into CO_2.

Excess air	**Moisture content Wt = 15%**	**Moisture content Wt = 68%**
α =1,60	1379	801
α=1,90	1217	728
α=2,20	1089	667
α=2,50	987	616
α=2,80	903	572
α=2,95	866	552
α=3,10	833	534

Table 4. Theoretical (adiabatic) combustion temperature of soybean straw (°C)

Also was made a bale straw moisture test on the basis of statistical data on the amount of rains on the territory of Belgrade in September, October and November. In September fell 3.9 kg rain/m^2 of soil, in October 98.8 kg rain/m^2, while in November fell 62 kg rain/ m^2. This means that the total per bale could fall about 386 kg of water, taking into account data from October and November. If we accept that the initial bale moisture was 10% and the average weight of bales after baling was about 200 kg by the calculation is to the point that one rain bales after October and November had a moisture content of ≈ 70%, which agreed quite well with the obtained moisture analysis of samples. So the increase of moisture in the bales stored in groups is a consequence of atmospheric rains.

Information about the theoretical combustion temperature (Table 4) confirmed the facts of bales combustion impossibility from the upper row of the crowd with such high moisture content in it. On the other hand, the vast majority of straw samples from bales, which were in the middle of the crowd (except than bale no. 2), has a moisture content ranging from 11.83 to

20.65%. This shows that such bales, with such quality, can satisfy needs of combustion in a boiler with a cigarette burning like that installed in PKB Corporation. This means that careful bales choosing avoid many problems in the boiler operation, would reduce the number of delays, to facilitate the boiler operators and that, most importantly, and would provide a safer production in the greenhouse that uses heat produced by combustion of soy bales straw.

We will point to another very important effect of increased baled biomass moisture, which is reflected in the flow, and thus indirectly on the kinetic characteristics of the biomass combustion in a boiler furnace. As is well known soybean straw baled is a porous medium, and as such unless the porosity is characterized by another feature of flow, which is permeability. Permeability is flowing layer in some fluid flow. Experiments have shown that as the permeability of porous biomass layer is less, the pressure drop i.e. flow resistance higher [24-26].

Bearing all this in mind, we can conclude that the bales moisture is essential for the applied concept of biomass combustion. During designing this boiler we calculated that the moisture content of bales shall not be greater than 25%. This is why we made this concept that the boiler is the simplest and cheapest solution for the user. For the case when one wants to burn fuel with a high content of moisture applied to the concept of boiler furnaces with higher volume and with additional support of liquid or gaseous fuel which satisfy necessary heat required for continuous combustion of fuel in the boiler furnace. It is more expensive and complicated option for users, both in design and manufacture of the boiler as well as its subsequent exploitation.

4.2. Experimental investigation in boiler furnace

Cigarette baled biomass combustion is a relatively new and unexplored technology. For this reason a complex Computation Fluid Dynamics (CFD) simulation of the combustion process at a specific procedure may be of importance for further investigation of the process of cigarette combustion. Numerical simulations process of this type of facility involves modeling the transfer of momentum, heat and substances during combustion of biomass bales, which composition is a porous medium [24, 27]. To form a mathematical model of thermo physical parameters except combustion in a porous medium, it is necessary knowledge inputs as thresholds model.

This paper describes experimental studies performed on mentioned boiler in order to determine the necessary model input parameters. When performing experiments measured the all parameters necessary to determine the global kinetics of the combustion process, the composition and temperature of flue gases at the outlet section of the space being modeled, and estimates the amount of fuel which is unburnt and which post combustion performed in a fluidized bed of its own ashes. In order to compare with the model made the determination or measurement of the temperature profile in soybean bale on its way from entering the combustion chamber to the combustion zone.

Experimental investigation on the demonstration boiler implies the measurements of the following input parameters [28]:

- mass flows of air at the entrance , $\dot{m}_1$, $\dot{m}_2$ and $\dot{m}_3$;

- temperature of the flue gas at the ash fluidized bed layer exiting T_{m3};

- mass flow of fuel (soybean straw baled $\dot{m}_4$);

- flue gas temperature at the exit cross section of model T_g;

- flue gas composition, at the exit intersection of model (CO_2, O_2, CO, NO).

Schematic of the experimental tests is shown in Figure 21. Measurement of mass flow of air at inlet cross sections was done indirectly through the measurement of velocity in the channels using Pitot-Prandtl's probe. Temperature of the flue gas exiting the fluidized bed of its own ashes, and the temperature at the outlet cross section were measured continuously by thermocouples type K. During experiments the acquisition of measurement data using corresponding instrument was performed.

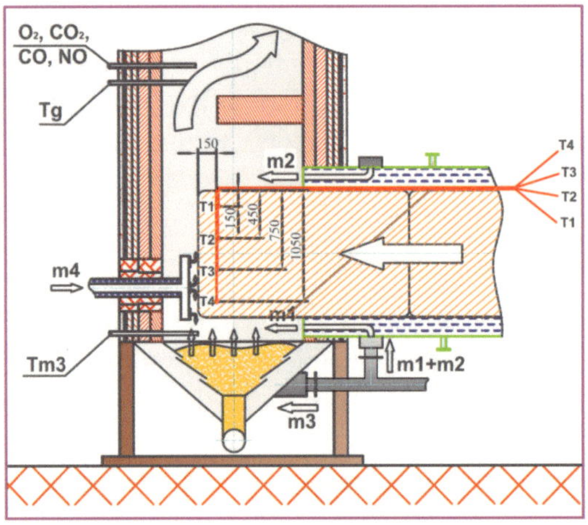

Figure 21. Measurements scheme at demonstrating hot water boiler furnace

Bales feeding were done discontinuous so that the cycle of 1 min the number of seconds the ball travels to the furnace and the rest to 1 min ball is at rest. Therefore, it is done recording the relative position of the bale in relation to entering the combustion chamber and time, which is based on data received on its secondary mass flow $\dot{m}_{fu}$. Composition of dry flue gas at the exit cross section was measured using a gas analyzer. Based on the measured air flow at the entrance to the ash layer $\dot{m}_{fu}$ and the temperature difference T_{m3} between inlet and outlet flue gases it is possible to determine the degree of conversion of coke residue, based on the energy balance between the energy of combustion of carbon, which falls on the fluidized bed and the enthalpy difference above the entrance of air and flue gas exit from the fluidized bed.

http://dx.doi.org/10.5772/51095

Also, were carried out experimental studies to determine the temperature profile in the central plane of soya straw bales, which participates in the combustion process. For this experiment, four thermocouples were placed in the central plane of height, according to the scheme at Figure 21. The experiment was performed in a stationary regime of the furnace operation and the temperature measured in function of the position of thermocouples. The appearance of soya straw bales with thermocouples placed in the median plane, just before entering the bale feeding system is shown in Figure 22. Experiments were performed at the maximum capacity of the furnace of 1.57 MW. Proximate and ultimate analysis of combusted soybean straw is provided in Table 5.

Ultimate analysis					Proximate analysis			
C [%]	H [%]	N [%]	O [%]	W [%]	Vol. [%]	C_{fix} [%]	A [%]	H_d [MJ/kg]
45.2	7.0	0.5	47.3	11.35	60.73	20.91	7.049	13.981

Table 5. Ultimate and proximate analysis of soy straw used in tests

4.3. Results of the experimental investigation

Experimental tests were carried out at a temperature in the furnace between 850-900°C, which is the optimum temperature for combustion of soybean straw. This temperature is high enough for complete combustion of straw, and safe from the point of ash melting. The stationary measurement regime remain several hours, but here will be presents the results of measurements for 1 h. It is enough to perform the necessary conclusions about the quality of combustion and comparisons with the proposed model. Data whose values are not changed during the experiment are shown in Table 6.

Figure 22. Thermocouples on the bale in feeding system

Temperature of air and fuel inputs (T_1, T_2, T_3, T_4, T_{fu}) has not changed during experiments, and for simplicity, adopted their size of 300 K, because any error will not have much impact on the accuracy of the results.

Variable	Air mass flow on inlet, [kg/s]				Fuel mass flow, [kg/s]
	Inlet 1	Inlet 2	Inlet 3	Inlet 4	
Value	0.2632	0.2263	0.2371	0.05	0.112143

Table 6. Data of experimental values

Flue gases temperatures at the outlet section of the model at input 3 were measured continuously during the experiments and their values in a representative period of time are shown in Figure 23. It can be seen that the average temperature of gases at the outlet section was 889°C, and temperature of flue gases at the entrance to model 3 was ~420°C.

Based on the known average temperature T_{m3}, air mass flow and air temperature at the inlet 3, it is possible determined the amount of fixed carbon which burn off in a fluidized bed of its own ashes, according to the methodology presented earlier. This amount essential represents a part of unburnt primary fuel that burns in the porous layer, and on the basis of this information can be concluded about the global kinetics of the process based on the mass flow of fuel and the degree of conversion. The mass flow of fictitious components of volatiles and fixed carbon burning in the porous layer, calculated according to the proposed methodology are shown in Table 7.

Variable	m, [kg/s]			
	C_3H_8	CO_2	H_2O	C_{fix}
Value	0,0166	0,0185	0,0323	0,0188

Table 7. Volatile and fixed carbon mass flow values

The global kinetics of the process is not only defined by the degree of conversion of coke residue, but also by the degree of conversion of combusted gases in volatiles. From that reason measuring of the concentration of components in dry flue gas at the exit cross section was performed (Figure 21). The values of the measured concentration of CO_2, O_2, CO and NO are shown in Figures 24 and 25 for a period of one hour.

From Figure 25 it can be seen that the concentration of nitrogen oxides is around 160 ppm, which is converted in mg/m^3 for the reference value of oxygen in the flue gas of 11% [29], is approximately 350 mg/m^3. The concentration of carbon monoxide was, at first view, very high, but we should bear in mind the fact that the observed cross section in which the measured concentration of the combustion process does not end, but it is continuing inside the chamber for burn of.

http://dx.doi.org/10.5772/51095

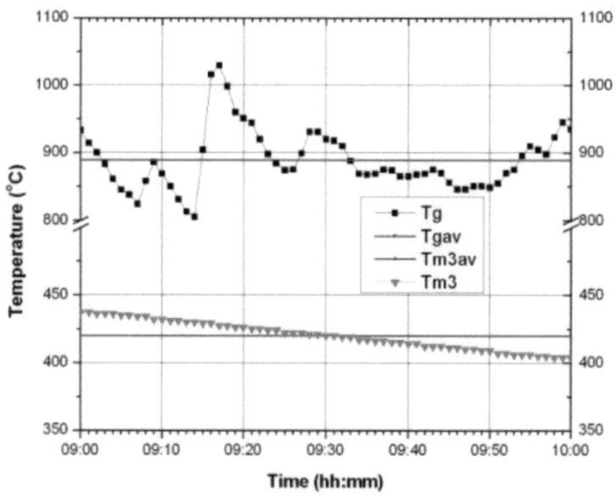

Figure 23. Flue gas temperature T_g and T_{m3} (T_{gav} = 889°C and T_{m3av} = 420°C)

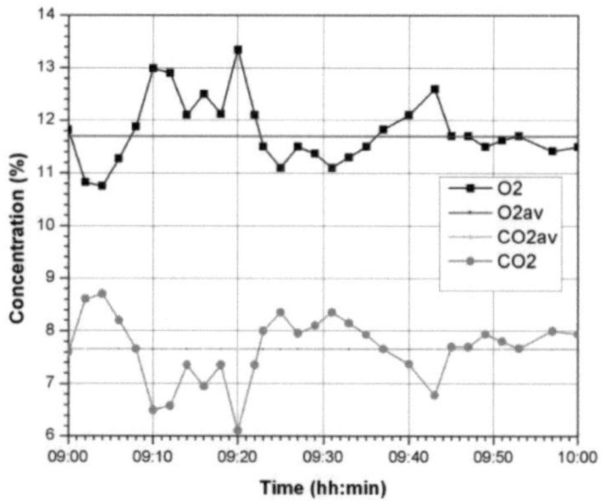

Figure 24. Carbon dioxide and oxygen concentration in dry flue gas on the outlet

The second part of the experimental research was related to determine the temperature field inside the soybean straw bale on its way from entering the furnace to the combustion zone. Graphical presentation of this measurement is shown in Figure 26.

It is important to note that the feeding rate of bale was not continual. In a determined time the bale was traveling to the combustion zone (working interval), and in determined period of time the bale was in pause (interval mode). Here it is clear that the average bale speed, if it is continuously moving, was less than the speed of movement in the working intervals. This statement is very important from the point of adoption of relevant temperature in the interval mode, because it is a case of unsteady heat conduction, so in the diagram can be seen more value of the temperature in one position. If the movement of the bale was uniformly and continuously, then the temperature is in a position that corresponds to the interval mode must have been between maximum and minimum measured values. Unfortunately, can not say with certainty whether the value was closer to the maximum, minimum or mean value. For the bale zero position we adopted the position of the combustion zone.

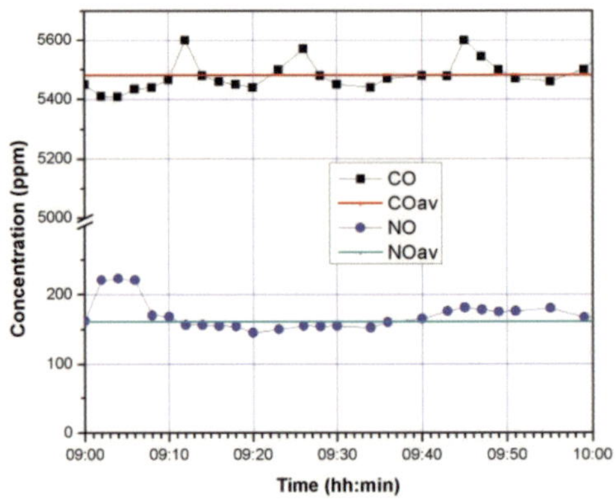

Figure 25. CO and NO concentration in dry flue gas on the outlet

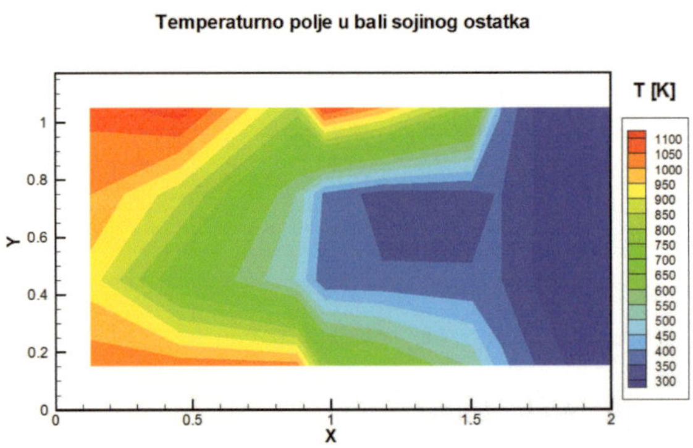

Figure 26. Measurement temperature profiles inside the soybean straw bale

4.4. Small scale plant for combined heat and power generation

The best way for utilizing residual agricultural biomass for energy production in industrial or district heating is to be used close to place of its gathering - in large agricultural companies. That is the optimal solution, from energy, as well as economy point of view. One of the most efficient ways, recommended by many institutions worldwide, is the combined heat and power (electricity) production - CHP, which use residual biomass as fuel, and have as least as possible power consumption. Proposed facility, analyzed here, meets those requirements in total. This is going to be the first CHP facility in Serbia, using residual agricultural biomass.

a) Present situation

Agricultural Corporation "Belgrade" - PKB is the largest agricultural enterprise in Serbia (≈ 22.000 ha of arable land), with agriculture and livestock as main domain of work. Crops (≈ 25.000 t/year of corn, wheat, barley, ≈ 5.760 t/year of soya, rapeseed and sugar beet), milk and vegetables are the main products. Each year after the harvest, a huge amount of soy straw (≈ 3.000 t/year) and corn stack (10-15.000 t/year) remain on the fields. The thermal facility with 1.5 MW power built for heating 1 ha greenhouses, has been working for five years already, using soy straw as main fuel. In the boiler, an original cigarette type combustion technology has been applied.

The energy efficiency of the straw utilization cycle is mainly affected by process of its preparation (balling, chopping, bundling), thus, the most justifiable is to use it close to the place of growing and gathering, and in form in which it is collected from fields. To utilize the remaining balled straw and meet thermal needs of surrounding objects, it is desirable to build a new boiler facility with an efficient and environmentally considerate combustion technology. The most efficient way is a combined heat and power (CHP) plant [30-32].

The planned facility would heat several objects: two greenhouses, the greenhouse office building, a school, and a hospital. An overview of the objects, their installed or required thermal capacity and current heating modus are given in Table 8.

Object	Area [m²]	Fuel	Annual fuel consumption/ Thermal power
Greenhouse 1	1 ha	Soy straw or light fuel oil	1.5 MW
Greenhouse 2	1 ha	No heating	1.5 MW
Greenhouse office building	130 [m²]	Soy straw	25 kW
School	4025 [m²]	Light fuel oil	110 t/year (600 kW)
Hospital	8600 [m²]	Heavy fuel oil	300 t/year (1.3 MW)

Table 8. Heating in Padinska Skela - present situation

b) Projected solution

The project consists of the substitution of existing boilers fed by fossil fuel by a CHP biomass facility, to heat public buildings and greenhouses as well as generate electricity. This will contribute to the reduction of CO_2 emissions and to the improvement of the general living conditions of the local inhabitants. As a pilot project, it has the potential to serve as an example for profitable green energy production facilities with replication potential.

The project covers a new boiler house with cogeneration facility. The boiler house comprise following main elements, included in the business plan (Table 9.). In addition to the mentioned units of the CHP facility, there are other infrastructural elements of the described technology, which have not been included in the business-plan, listed in Table 10.

The planned CHP facility is based on the two proven technologies:

a. The balled straw combustion technology (developed in Laboratory for Thermal Engineering and Energy of Vinca Institute of Nuclear Sciences, in cooperation with Company Tipo-Kotlogradnja, Belgrade), which has been applied in existing 1.5 MW facility used for heating greenhouses in PKB.

b. Organic Rankin Cycle (ORC) for electricity generation. ORC technology is based on turbines driven by silicone oil steam (although, steam of other liquids can be used).

Functional scheme of the CHP facility with the proposed combustion technology and the heat into electricity energy conversion technology is shown in Figure 27. The combustion technology (based on the cigarette type burning) has been described in detail in numerous papers [9, 10, 20], and has been developed up to industrial application, largely with help of Ministry of Education and Sciences of Republic of Serbia. Beside the cigarette combustion, this technology comprises some original technical solutions, considering the organization of biomass combustion completion in fluidized bed. This technology enables using balled agricultural residues (in form it has been collected on fields), with no additional transformation (chopping, grounding), which decreases fuel cost and pollution, and contributes to energy efficiency. This type of combustion has been marked as the most suitable way of combustion of agricultural biomass in EU [13]. This technology, proven in operation in PKB, is going to be applied for CHP facility and the supplemental (reserve) biomass boiler.

The cogeneration technology is ORC - Organic Rankin Cycle based. There are a number of well known producers of the equipment. The technology is used exclusively for power generation in CHP facilities, with electric efficiency is up to 20% and overall efficiency of over 80%. More than 100 of facilities like that have been installed in EU, thus the technology could be considered proven.

Combining these two technologies would give the first experimental, demonstrational and industrial facility of the kind. It is going to be a good reference for all companies involved in the project. The facility is going to be built in PKB Company, and the company is going to supply it with the fuel (soy straw, rapeseed straw and cornstalk), which is a very convenient because they have an experience in operation with a similar boiler. Public Company "Belgrade Plants" also owned by the city of Belgrade, should take part in design and construction of the necessary infrastructure (connecting the existing system with the new one). It is necessary to note that a heat accumulator (hot water reservoir) is going to be built in scope of the boiler house in order to cover the peaks in energy consumption. It has been proven in operation with the existing cigarette type boiler in the PKB.

Boiler house element	Preliminary data and description
CHP biomass facility with the auxiliary equipment	3000-4000 kW thermal + 400-600 kW electric (net)
Hot water heavy fuel oil (or light fuel oil) boiler with the whole auxiliary equipment	3000-4000 kW (as a reserve)
Heat accumulator	Matching the peaks in energy consumption
The boiler house building (the new boiler house is going to be connected to the existing one)	Building for the accommodation of the equipment
Reconstruction of the existing boiler house and connection it with the new one	Connecting the existing and new boiler house into a functional unit
Automated bale storage and equipment for the bale manipulation	Next to the boiler house, week bale storage is planned

Table 9. The main project elements

Infrastructural element	Preliminary data and description
Pipeline for connecting the consumers with the new boiler house	≈1000 m pipeline
Heating installations in the new greenhouse	Same heating system as in the existing greenhouse
Connecting the facility to the electric grid	Installation of necessary electric poles and equipment
Communicational roads	Hard material roads
Building of a central roofed storage of biomass for annual needs	The existing storage does not meet present requirements
Machinery for gathering and transport of the biomass	Machinery for provisioning the biomass reserves in optimal conditions

Table 10. Infrastructural elements, not comprised by the business-plan

c) *Project Implementation*

Construction project subjected to the CHP plant, in the suburb of Padinska Skela, near Belgrade, presents the continuing efforts from Laboratory for thermal engineering and energy of Vinca Institute and Central European Initiative (CEI) to build such a plant in Serbia. CEI associates through implementation of projects BIOMADRIA and BIOMADRIA 2 recognized the importance of the construction of such a facility, as similar projects can be transferred to many places in Serbia, as well as in the surrounding countries, which also has an intensive agricultural production.

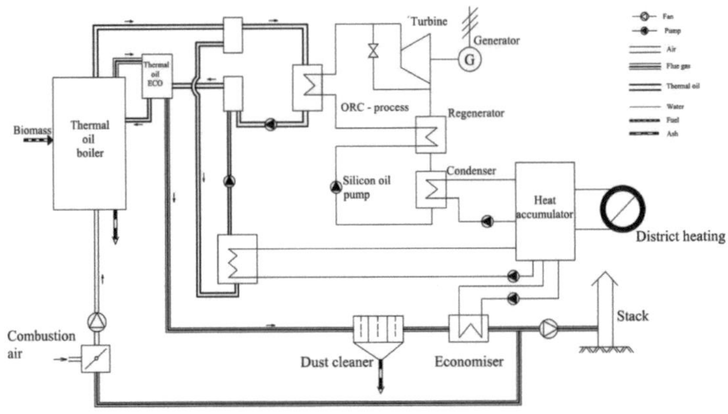

Scheme of the CHP facility in Padinska Skela

Figure 27. Functional scheme of the combined heat and power facility

The time needed for the full implementation of the project is 1.5 years starting from the date of loan approval and the first disbursement (6 months for preparations, obtaining all necessary permissions and licenses, and designing; 8 months for the building the facility, and 4 months for its commissioning). The service lifetime of the CHP plant would be 25 years, which is a commonly accepted lifetime with proper maintenance.
d) CHP facility parameters

Plant parameters are determined on basis of heat demand. In this case, the analysis is complicated by different heating dynamics of the consumers (greenhouses, office building, school, hospital). The school has a different heating dynamic compared to the hospital, and all this is completely different from the dynamics of greenhouse heating system, e.g. the greenhouse needs heating at night while hospitals and schools are heated during the day. The consumers' heat demand has been carried out in the three steps:

a. Analysis of the heat demand in the objects where people are staying: it's been carried out on the basis of data on average monthly fuel consumption, provided by Public Company "Belgrade Plants" (October and April 5%, November 11%, March 14%, December and February 20% and January 25%).

b. Analysis of the heat demand in the greenhouses: carried out on the basis of the heat demand, as well as the plants which are grown in the greenhouses. The data are averaged for a three-year period, and it show that average fuel consumption are varied between 4% in September up to 20% in January and April.

c. Analysis of the needed active power during the heating season: In order to calculate optimal power of the planned facility, the analysis of the minimal, maximal and average heat demand have been carried out for each object, as well as the overall calculation. After that, the installed thermal power of the facility can be established. The optimal solution is the one which demands minimal investment, with maximal potential gain. Analysis showed that average annual active power in the heating season is 45% of the installed power for the hospital, 38% for the school, 35% for the office building and 55% for the greenhouses.

5. Conclusions

Energy potential of renewable energy sources in the Republic of Serbia equals approximately 4.3 Mtoe/year. Biomass is deemed to be the main source of renewable energy, with estimated 2.7 Mtoe/year of energy potential, with 60% being the potential of agricultural biomass and the remaining 40% being attributed to the forest biomass. Combustion of agricultural biomass was analyzed with respect to the cigar burner combustion technology, suitable for the whole-bale combustion. Technology developed was tested in the 75 kWth hot water boiler, 1 MWth demonstration furnace and 1.5 MW industrial hot water boiler, where combustion of soybean and rapeseed straw samples was investigated. Results obtained indicated that combustion technology developed was very convenient for combustion of biomass varieties characterized by high ash melting temperatures. A cigar firing combustion system is expected to exhibit the following advantageous features: a) combustion of whole bales and whole energy crops; b) compact combustor design; c) short start up period, good load-following performance; d) profitable operation of smaller facilities (down to 1 MWth); e) division of combustion from the heat recovery system, usable not only for the provision of steam (for heat generation or CHP), but also as a hot gas generator in industrial drying applications.

Cigar burner combustion system promises a more competitive use of renewable for "green" heat and power generation as well as their use in various industrial applications. In the same time, biomass combustion in cigar burners was modeled by appropriately developed numerical model. The model developed enabled the effect of fuel moisture content on the temperature distribution in the furnace to be analyzed, as well as related emissions of harmful combustion products into the environment. Research investigation conducted has demonstrated that high combustion temperatures can be achieved in furnaces used for the combustion of agricultural biomass and that achieved CO and NOx emission levels are lower than the regulatory emission limit values defined by Serbian legislation.

Acknowledgements

The paper has been realized in scope of Ministry of Education and Science of Republic of Serbia's project "Development and improvement of technologies for energy efficient and environmentally sound use of several types of agricultural and forest biomass and possible utilization for cogeneration", Record number III42011.

Author details

Branislav S. Repić[1*], Dragoljub V. Dakić[2], Aleksandar M. Erić[1], Dejan M. Đurović[1], Stevan D. J. Nemoda[1] and Milica R. Mladenović[1]

*Address all correspondence to: brepic@vinca.rs

1 Vinca Institute of Nuclear Sciences, Laboratory for Thermal Engineering and Energy, University of Belgrade, P.O. Box 522, 11001 Belgrade, Serbia

2 Innovation Center, Faculty of Mechanical Engineering, University of Belgrade, Kraljice Marije 16, 11120 Belgrade, Serbia

References

[1] Government of Republic of Serbia. (2010). Decree on amendments and supplements to the decree on program for the realization of the energy sector development strategy of the Republic of Serbia until 2015 for the period 2007-2012. *Official Gazette of Republic of Serbia* [27].

[2] The Republic of Serbia Biomass Action Plan for The Republic of Serbia 2010-2012. Ministry of infrastructure and energy and NL Agency, The Netherlands, Serbian-Dutch Government to Government Project on Biomass and Biofuels G2G08/SB/6/3, Belgrade 2010.

[3] Dodić, S., Zekić, V., Rodić, V., Tica, N., Dodić, J., & Popov, S. (2010). Situation and perspectives of waste biomass application as energy source in Serbia. *Renewable and Sustainable Energy Reviews*, 14, 3171-3177.

[4] De Wit, M., & Faaij, A. (2010). European biomass resource potential and costs. *Biomass and Bioenergy*, 34, 188-202.

[5] Repic, B., Dakic, D., Djurovic, D., & Eric, A. (2010). Development of a boiler for small straw bales combustion. *Nathwani J, Ng AW, editors. Paths to sustainable energy*, Rijeka, InTech, 647-664.

[6] http://www.mie.gov.rs Ministry of infrastructure and energy of Republic of Serbia.

[7] Saidur, R., Abdelaziz, E. A., Demirbas, A., Hossain, M. S., & Mekhilef, S. (2011). A review on biomass as fuel for boilers. *Renewable and Sustainable Energy Reviews*, 15, 2262-2289.

[8] Repić, B., Dakić, D., Djurović, D., Erić, A., & Paprika, M. Small scale plant for combined heat and power generation utilizing local biomass. In: SIMTERM2011. *Proceedings of the 15th Symposium on Thermal Science and Engineering of Serbia; 2011Oct 18-21; Sokobanja, Serbia*, 283-292.

[9] Repić, B., Dakić, D., Paprika, M., Mladenović, R., & Erić, A. (2008). Soya straw bales combustion in high efficient boiler. *Thermal Science*, 12, 51-60.

[10] Mladenović, R., Dakić, D., Erić, A., Mladenović, M., Paprika, M., & Repić, B. (2009). The boiler concept for combustion of large soya straw bales. *Energy*, 34, 715-723.

[11] Repić, B., Dakić, D., Djurović, D., & Erić, A. The cigar burner combustion system for baled biomass. CHISA2010. CD ROM Proceedings of the 19th International Congress of Chemical and Process Engineering and 7th European Congress of Chemical Engineering ECCE-7. 2010Aug 28 Sep 1; Prague, Czech Republic. Paper 978-8-00202-210-7 , I6(3), 1-12.

[12] Quaak, P., Knoef, H., & Stassen, H. (1999). Energy from biomass- A review of combustion and gasification technologies. *World Bank technical paper 422 Energy series, Washington, USA.*

[13] Kavalov, B., & Peteves, S. D. (2004). Bioheat applications in the European Union: An analysis and perspective for 2010. *European Commission, Directorate-General Joint Research Centre, Institute for Energy.*

[14] Repić, B., Dakić, D., Mladenović, R., & Erić, A. Development of a boiler based on cigarette combustion principle for straw bales from agricultural production. *CHISA2008 CD ROM Proceedings of the 18th International Congress of Chemical and Process Engineering, 2008 Aug 24-28, Praha, Czech Republic, Paper,* P5(264), 1-12, 978-8-00202-047-9.

[15] Repić, B., Dakić, D., Paprika, M., Mladenović, R., & Erić, A. An efficient boiler burning small soya straw bales, In: Guzović Z, Duić N, Ban M, editors. CD ROM Proceedings of the 4th Dubrovnik Conference on Sustainable Development of energy, Water and Environment Systems, 2007 June 4-8, Dubrovnik, Croatia, 13978953631387 , 1-8.

[16] Dakić, D., Repić, B., Erić, A., Djurović, D., & Paprika, M. (2011). Development of small balle feeding devices for use in agricultural biomass combustion systems. *Contemporary Agricultural Engineering*, 37, 165-174.

[17] Kraus, U. Test results from pilot plants for firing wood and straw in the Federal Republic of Germany. *Palz W, Coombs J, Hall DO, editors. Proceedings of the 3rd E.C. Conference Energy from Biomass, 1985 Elsevier Applied Science Publishers, London*, 799-803.

[18] Dakić, D., Paprika, M., Mladenović, R., Erić, A., & Repić, B. The boiler concept for combustion of large soya strow bales as the zero emission plant. In: Guzović Z, Duić N, Ban M, editors. *CD ROM Proceedings of yhe 4th Dubrovnik Conference on Sustainable Development of Energy, Water and Environment Systems, 2007 June 4-8, Dubrovnik, Croatia*, 1-11, 13978953631387.

[19] Mladenović, R., Erić, A., Mladenović, M., Repić, B., & Dakić, D. *Energy production facilities of original concept for combustion of soya straw bales. In: CD ROM Proceedings of the 16 European biomass conference & exibition „From research to industry and markets", 2008 June 2-6, Valensia, Spain*, 1260-1270, 978-8-88940-758-1.

[20] Turanjanin, V., Djurović, D., Dakić, D., Erić, A., & Repić, B. (2010). Development of the boiler for combustion of agricultural biomass by products. *Thermal Science*, 14, 707-714.

[21] Djurović, D., Dakić, D., Erić, A., & Repić, B. *Development of the boiler for combustion of agricultural biomass by products. In: Guzović Z, Duić N, Ban M, editors. CD ROM Proceedings of the 5th Dubrovnik Conference on Sustainable Development of Energy, Water and Environment Systems, 2009Sep. 29th-Oct. 3th, Dubrovnik, Croatia,* 1-18, 978-9-53631-398-3.

[22] Dakić, D., Erić, A., Djurović, D., Erić, M., Živković, G., Repić, B., Mladenović, M., Nemoda, S., Mirkov, N., & Stojanović, A. *One approach of using the agricultural biomass for heating. In: Spitzer J, at all., editors. Proceedings of the 18th European Biomass Conference "From research to industry and markets", 2010 May 3-7, Lion, France,* 1944-1948, 978-8-88940-756-5.

[23] www.hidmet.gov.rs Republic Hydrometeorological Service of Serbia.

[24] Erić, A. (2010). Thermomechanical Processes During Baled Soya Residue Combustion in Pushing Furnace. PhD thesis. Faculty of Mechanical Engineering University of Belgrade. *Belgrade.*

[25] Erić, A., Dakić, D., Nemoda, S., Komatina, M., & Repić, B. (2011). Experimental method for determining Forchheimer equation coefficients related to flow of air through the bales of soy straw. *International Journal of heat and mass transfer*, 54, 4300-4306.

[26] Erić, A., Dakić, D., Nemoda, S., Komatina, M., & Repić, B. (2010). Determination of stagnant thermal conductivity coefficient for baled soy bean residue. *Contemporary Agricultural Engineering*, 36, 334-343.

[27] Djurović, D. (2011). Combustion of the gases in adiabatic furnace for agricultural biomass combustion. PhD thesis. Faculty of Mechanical Engineering University in Belgrade. *Belgrade.*

[28] Erić, A., Dakić, D., Nemoda, S., Komatina, M., Repić, B., Mladenović, M., & Djurović, D. (2011). Experimental investigation of baled soy straw combustion in cigarette furnace. *Contemporary Agricultural Engineering*, 37, 153-164.

[29] Regulation on the emission limit values of the pollutants in the air. (2010). *Official Gazette of Republic of Serbia* [71].

[30] Djurović, D., Dakić, D., Repić, B., Nemoda, S., Živković, G., & Erić, A. (2010). Economic feasibility of building a facility for combined heat and power production using agricultural biomass. *Contemporary Agricultural Engineering*, 36, 373-381.

[31] Repić, B., Dakić, D., Janić, T., Djurović, D., & Erić, A. (2011). Economic feasibility of building plant for heat energy production in the slaughter-house industry using biomass. *Contemporary Agricultural Engineering*, 37, 145-152.

http://dx.doi.org/10.5772/51095

[32] Government of Republic of Serbia, Decree on incentive measures for electricity generation using renewable energy sources and for combined heat and power (CHP) generation. (2009). *Official Gazette of Republic of Serbia* [99].

An Optimized Maximum Power Point Tracking Method Based on PV Surface Temperature Measurement

Roberto Francisco Coelho and Denizar Cruz Martins

Additional information is available at the end of the chapter

http://dx.doi.org/10.5772/51020

1. Introduction

On the last decades, distributed generation (DG) systems based on photovoltaic (PV) generation are slowly been introduced to the world energy matrix, in which some important aspects as political incentive, cost reduction, electricity rising demand, improvements on PV materials and increasing on power converters efficiency have contributed to the present scenario [1-3].

From the power processing point of view, high efficiency conversion, by itself, cannot ensure the optimized power flow, since the PV output voltage and current are strongly dependent on environmental conditions, i.e., solar radiation and temperature; however, on the literature, many works bring solutions to maximize the photovoltaic output power, employing specific circuits denominated by Maximum Power Point Trackers (MPPT) [4-8]. In most applications, the MPPT is a simple dc-dc converter interposed between the photovoltaic modules and the load, and its control is achieved through a tracking algorithm.

The studies on the MPPT area are normally grouped in two categories: the first one relates to the dc-dc converter topology optimization, focusing on methods to determine a suitable dc-dc converter for operating as MPPT [9]; and the second one refers to the maximum power point tracking algorithm, responsible for properly controlling the dc-dc converter in order to establish the system operating point as close as possible to the Maximum Power Point (MPP). Therefore, an *efficient* MPPT system need to be composed by the integration of an adequate dc-dc converter (hardware) and proper tracking algorithm (software), resulting in some desired aspects:

- Fast tracking response (dynamic analysis);
- Accuracy and no oscillation around the MPP (stead-state analysis);

- Capacity to track the MPP for wide ranges of solar radiation and temperature;
- Simplicity of implementation;
- Low cost.

The most popular algorithms employed in PV tracking systems [10-18] - Constant Voltage, Perturb and Observe (P&O) and Incremental Conductance (IncCond) – are extensively explored by specialized literature, nevertheless, since fast tracking response and accuracy conflict one from other, the mentioned tracking methods cannot satisfy, simultaneously, both of them. In place of the traditional and spread methods, some authors have proposed complex MPPT algorithms, based on fuzzy logic and neural network, in order to accomplish fast tracking response and accuracy in a single system. These proposals, nevertheless, present some disadvantageous: needed for high processing capacity, complexity, cost elevation and, in some cases, employment of extra sensors.

In this chapter, PV maximum power point tracking systems are analyzed under two distinct points of view: firstly, the influence of the dc-dc converter on the tracking quality is accounted. In this study, the effect of solar radiation, temperature and load variations are considered, and the tracking performance of Buck, Boost, Buck-Boost, Cuk, SEPIC and Zeta converters are compared. Secondly, a new tracking algorithm, based on the PV surface temperature, is introduced. The advantages concerning the proposed method come from the simplicity, low cost, analogue or digital implementation, fast tracking response, accuracy and no oscillation around the MPP on steady state.

In order to achieve the main chapter topics, a brief revision of PV generation is highlighted in next section.

2. Photovoltaic Generation

Photovoltaic modules output power depends on environmental conditions, such as solar radiation and temperature, resulting in a non-liner and time-variant power source. The employment of a PV generator only can be successfully attained if it is correctly characterized.

2.1. Relationship Among PV Cell, Module and Array

Photovoltaic cells are the basic building blocks on construction of PV power systems. The amount of power delivered by a PV cell is, typically, restricted to few Watts, due to the surface area limitation. For raising the generated power, in order to reach hundreds of Watts, PV cells may be grouped in a PV module. Similarly, it is possible to connect a group of PV modules (series, parallel or both) in order to obtain a PV array, whose power range is established from kilo-Watts to mega-Watts [19]. The distinction among PV cell, module and array is illustrated at Figure 1.

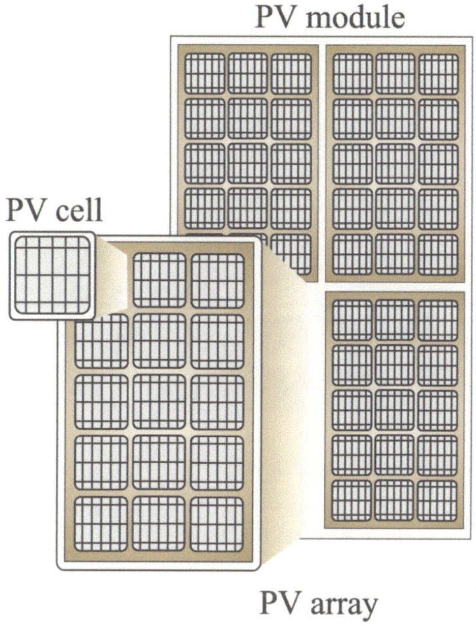

Figure 1. PV array composed by an arrangement of PV modules and PV module composed by an arrangement of PV cells.

2.1.1. *Standard Test Condition*

The Standard Test Conditions (STC) refers to the conditions under which PV modules are tested in laboratory. STC defines the values of irradiance, temperature and air mass index, in which the manufactures feature the PV devices, permitting to compare their performance and efficiency conversion.

2.1.2. *Irradiance*

The Sun energy reaches to the Earth through electromagnetic waves, resulting in an irradiance (or solar radiation) of about 1366 W/m^2 on its outer atmosphere. However, due to atmospheric effects – scattering, absorption and reflection -, the incoming irradiance is modified before reaching the Earth's surface [20].

The process of scattering occurs when small particles and gas molecules diffuse the radiation in random directions, while absorption is defined as a process in which the solar radiation is retained by atmosphere substances and converted into heat. In addition, part of the total solar radiation is redirected back to the space by reflection and part, termed by direct solar radiation, reaches the Earth's surface unmodified by any of the above atmospheric processes, as depicts Figure 2.

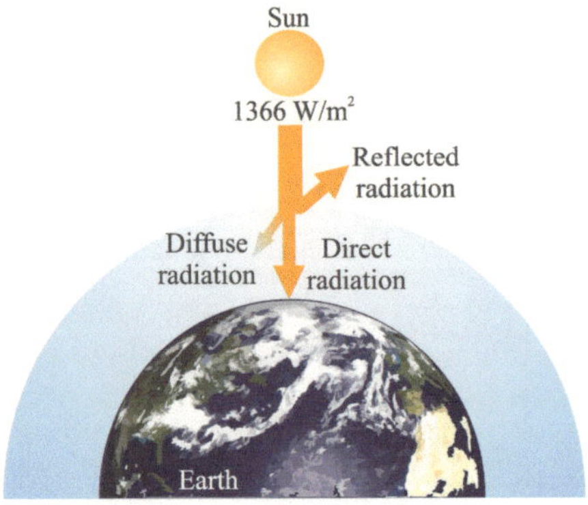

Figure 2. Atmosphere effects on incoming solar radiation.

Since the direct radiation on a clear day, at noon, is typically 1000 W/m^2, this value is adopted as reference at STC.

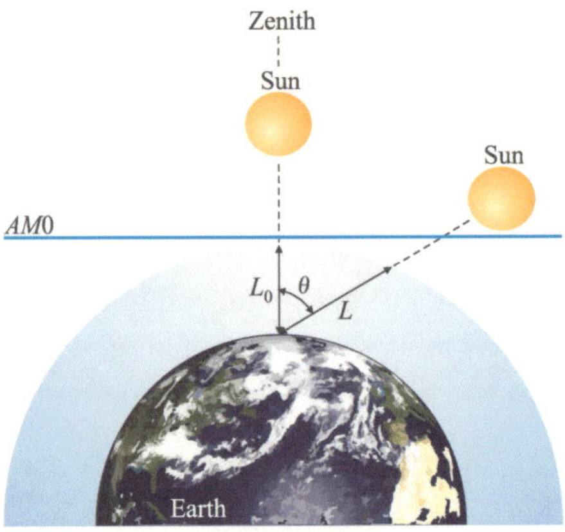

Figure 3. Solar radiation path across the Earth's atmosphere.

2.1.3. *Air Mass*

The Air Mass (*AM*) index quantify the solar radiation path (*L*) across the atmosphere, normalized by the shorter path (L_0), measured from the zenith angle, as is depicted in Figure 3 and mathematically described by (1) [19]. The index *AM*0 is used to describe the radiation path out of the atmosphere, where the irradiance is constant and equal to 1366 W/m^2.

$$AM = \frac{L}{L_0} = \frac{1}{\cos(\theta)} \tag{1}$$

On industry, PV modules are standardized considering an air mass index of AM=1.5. This value comes from a angle of aproximatly 48°, proper representing the PV instalitions around the most populed centres across Europe, China, Japan and United States of America, located in mid-latitudes. This value is also adopted as reference at STC.

2.1.4. Temperature

The third parameter used to characterize a PV module is its surface temperature. The STC temperature value adopted for PV modules characterization is T=25 °C.

2.2. I-V and P-V generation curves

Under the specified Standard Test Conditions, expressed by (2), PV modules are tested and featured by I-V (current versus voltage) and P-V (power versus voltage) curves, in which the effect of solar radiation and temperature on the PV generated power is evidenced.

$$\begin{aligned} S^{STC} &= 1000\ \text{W/m}^2 \\ T^{STC} &= 25\ ^\circ\text{C} \\ AM^{STC} &= 1.5 \end{aligned} \tag{2}$$

Although both, solar radiation and temperature, are strongly coupled, solar radiation predominantly influences the PV module output current, while temperature mainly changes the PV module output voltage, as depicts the I-V curve presented at Figure 4, obtained from Kyocera KC200GT PV module datasheet.

One of most important PV module operation point is obtained on the knee of the I-V curve. In this point, named by maximum power point (MPP), the product of the PV output voltage and current results at the maximum available power, for a given solar radiation and temperature. For emphasizing the maximum power point, an alternative P-V (power versus voltage) curve may be plotted, in accordance with Figure 5.

Mathematical expressions for calculating the PV module output voltage V_{mpp} and current I_{mpp} on MPP are given by (3) and (4), whose product results on the PV output power P_{mpp}, according to (5) [21], where:

V_{mpp}, I_{mpp}, P_{mpp} : PV module output voltage, current and power on MPP for any irradiance (S) and surface temperature (T);

V_{mpp}^{STC}, I_{mpp}^{STC}, P_{mpp}^{STC} : PV module output voltage, current and power on MPP. These parameters are obtained from datasheet and specified on STC (T^{STC} and S^{STC});

μ_V, μ_A : Temperature voltage coefficient (V/°C) and temperature current coefficient (A/°C). These parameters are also obtained from datasheet.

$$V_{mpp} = V_{mpp}^{STC} + \left(T - T^{STC}\right)\mu_V \tag{3}$$

$$I_{mpp} = \frac{S}{S^{STC}} I_{mpp}^{STC} + \left(T - T^{STC}\right)\mu_A \tag{4}$$

$$P_{mpp} = \frac{S}{S^{STC}} P_{mpp}^{STC} + \left(T - T^{STC}\right)\left(\frac{S}{S^{STC}}\mu_V I_{mpp}^{STC} + \mu_A V_{mpp}^{STC}\right) + \left(T - T^{STC}\right)^2 \mu_V \mu_A \tag{5}$$

Note that (5) is useful once it allows estimating the amount of available PV power only from the environmental data (*S*, *T*). All other related parameters are commonly specified on PV module datasheet. For instance, considering the KC200GT PV module, the following datasheet specifications are found:

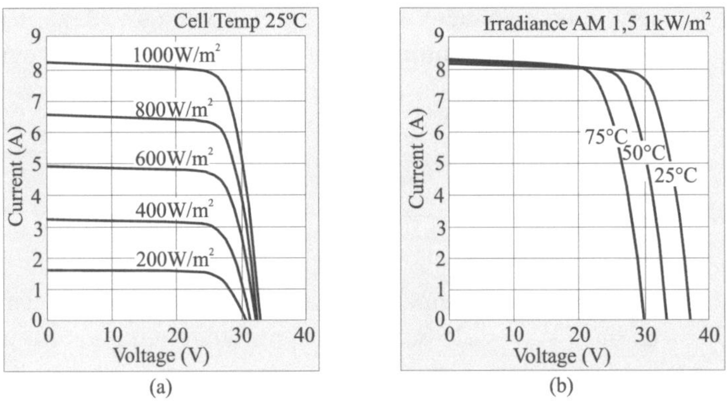

Figure 4. I-V curve from Kyocera KC200GT PV module: (a) under constant temperature and (b) under constant irradiance.

PV specified parameter	Value
V_{mpp}^{STC}	26.3 V
I_{mpp}^{STC}	7.61 A
P_{mpp}^{STC}	200 W
μ_V	-0.14 V/°C
μ_A	0.00318 A/°C

Table 1. PV module specifications on STC from Kyocera KC200GT PV module datasheet.

Furthermore, short circuit current (I_{sc}) and open circuit voltage (V_{oc}) are also important for a complete PV module characterization. They remark the points where the PV generated power is null, but the output current or voltage reach the maximum value, respectively. Figure 6 highlights I_{sc}, V_{oc}, I_{mpp}, V_{mpp} and P_{mpp} on both, I-V and P-V curves.

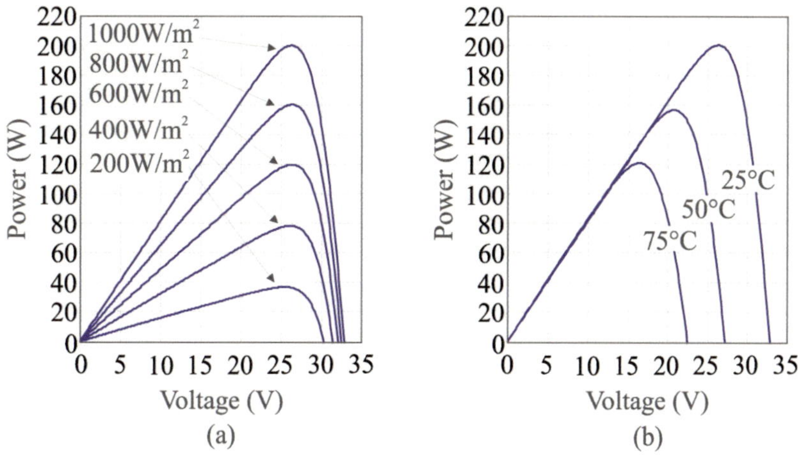

Figure 5. P-V curve from Kyocera KC200GT PV module obtained by simulation: (a) under constant temperature and (b) under constant solar radiation.

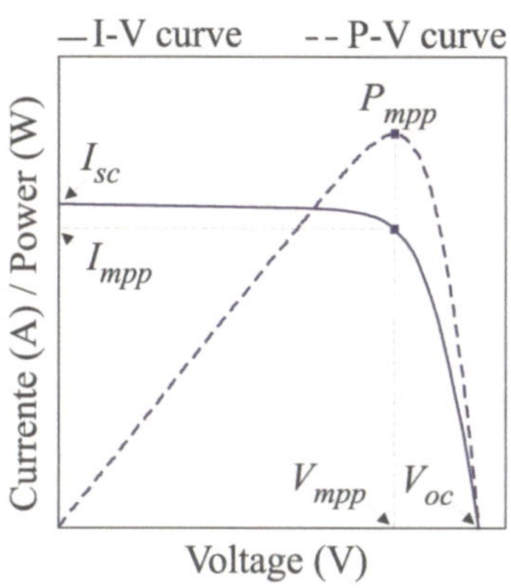

Figure 6. Identification of I_{sc}, V_{oc}, I_{mpp}, V_{mpp} and P_{mpp} on the I-V and P-V curves.

3. Maximum power point tracker

For maximizing the PV conversion efficiency, the incoming sun energy must be converted to electricity with the highest efficiency, accomplished when the photovoltaic module operates on the maximum power point. Nevertheless, since this operating point is strongly affected by the solar radiation and temperature levels, it may randomly vary along the I-V plan, as illustrates Figure 7.

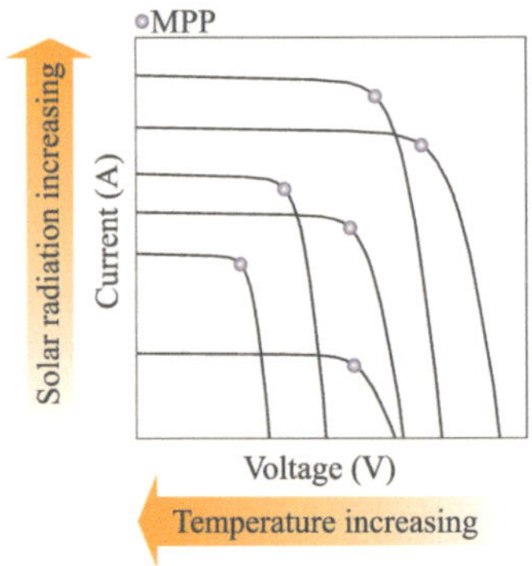

Figure 7. MPP across the I-V plan considering solar radiation and temperature changes.

Thus, in order to dynamically set the MPP as operation point for a wide range of solar radiation and temperature, specific circuits, known at the literature by Maximum Power Point Trackers (MPPT), are employed.

In this chapter the studies concerning MPPT are grouped in two categories: the first is related to hardware, in which the influence of the dc-dc converter and load-type on the tracking quality is investigated, and the second refers to the software, where tracking accuracy and speed are targeted.

3.1. MPPT from the dc-dc converter point of view

The operating point of a photovoltaic system is defined by the I-V generation and load curves intersection. For understanding how it occurs, firstly considerer a PV module supplying a resistive load, as depicts Figure 8.

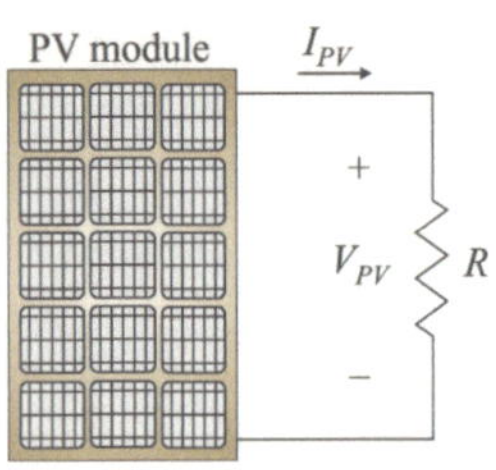

Figure 8. PV module supplying a resistive load.

The load curve is accomplished by the Ohm's Law, in accordance with (6), while the generation curve is related to the PV I-V curve. Both curves are represented at Figure 9.

http://dx.doi.org/10.5772/51020

$$I_{PV} = \frac{V_{PV}}{R} \tag{6}$$

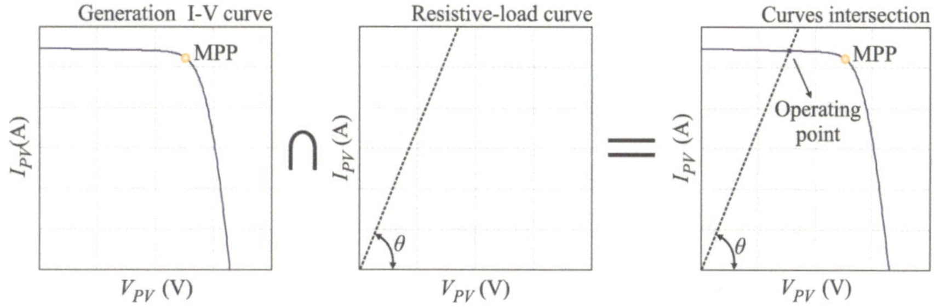

Figure 9. Definition of the system operating point by the I-V and load curves intersection.

Even when the load resistance is chosen for both curves intercept each other exactly on the MPP, it is impossible to ensure the maximum power transfer for long time intervals, once when solar radiation or temperature change, the MPP is relocated on the I-V plan.

For solving this problem, in order to maintain the system always operating on the MPP, the load curve should be modified according to solar radiation or temperature changes. For example, from Figure 10, if the PV generation curve is *I-V 1* and the load curve is *Load 1*, the system operating point is given by *MPP 1*. Now, considering a solar radiation and temperature change, the generation curve comes from *I-V 1* to *I-V 2*. In this situation, keeping the same load curve (*Load 1*), the system operating point is established at *X2*, i.e., out of the MPP. However, if the load curve is modified from *Load 1* to *Load 2*, the system backs to operate on the MPP, in this case, *MPP 2*.

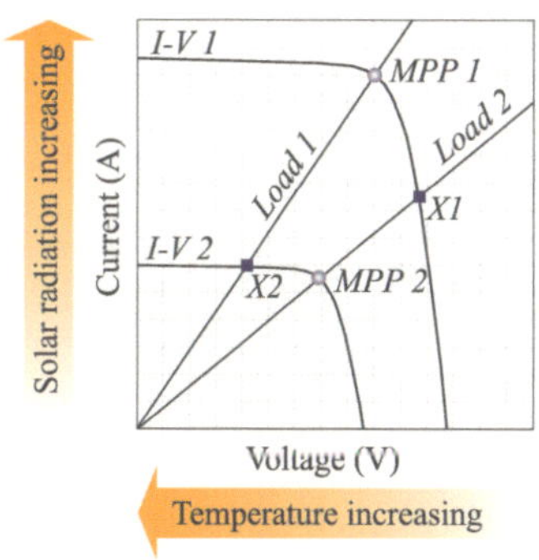

Figure 10. I-V and load curves intersection for defining the PV system operating point.

Evidently, modifying the load curve in accordance with the solar radiation and temperature changes is not a suitable solution, since the load is defined by the user. Nevertheless, if a dc-dc converter is interposed between the PV module and the load, it is possible to control the converter duty cycle in order to emulate a variable load from the PV terminals point of view, even when a fixed load is employed. The arrangement presented at Figure 11, composed by a PV module, a dc-dc converter and a load, defines the hardware of a maximum power point tracking system.

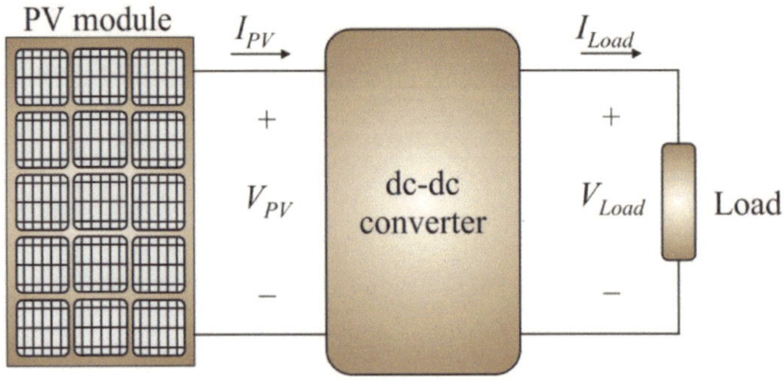

Figure 11. Maximum point tracker system.

It is important to emphasize that the tracking system will present distinct behaviours depending on the dc-dc converter and load-type features. Here, buck, buck-boost, boost, Cuk, SEPIC and zeta converters will be analyzed in association with resistive or constant voltage loads-type.

3.1.1. Analysis for resistive load-type

When a resistive load is connected to the dc-dc converter, Figure 11 may be redrawn as per Figure 12 and equation (7) can be derived.

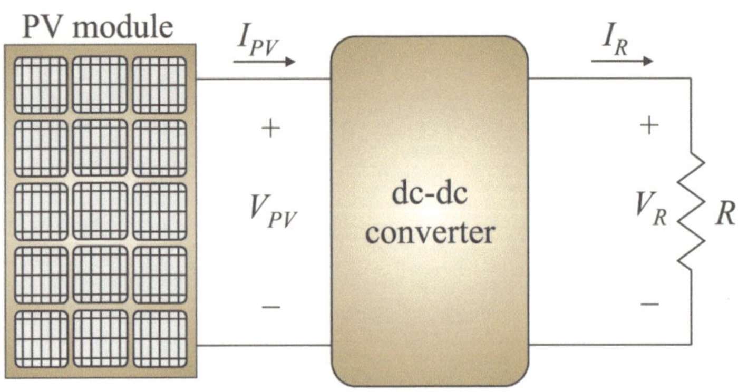

Figure 12. MPPT supplying a resistive load-type.

$$V_R = R I_R \tag{7}$$

Taking into account a literal dc-dc converters static gain G, the input system variables (V_{PV} and I_{PV}) can strictly be associated to the output ones (V_R and I_R), through (8) and (9).

$$G = \frac{V_R}{V_{PV}} \tag{8}$$

$$G = \frac{I_{PV}}{I_R} \tag{9}$$

Isolating V_R in (8) and I_R in (9) and substituting the found results into (7), it is possible to obtain (10).

$$\frac{V_{PV}}{I_{PV}} = \frac{R}{G^2} \tag{10}$$

The term V_{PV}/I_{PV} describes the effective resistance R_{eff} obtained from the PV module terminals. In other words, the dc-dc converter emulates a variable resistance, whose value can be modulated in function of the converter static gain G. This conclusion allows redesigning Figure 12 as Figure 13 and writes (11).

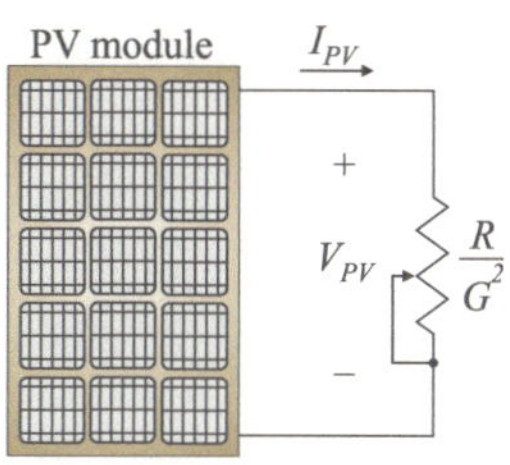

Figure 13. Effective resistance obtained from the PV module terminals.

$$V_{PV} = \frac{R}{G^2} I_{PV} \tag{11}$$

When plotted on the I-V plan, equation (11) results in a straight line whose inclination angle θ, given by 12, is modified according to the converter static gain G.

$$\theta = \operatorname{atan}\left(\frac{G^2}{R}\right) \tag{12}$$

Table 2 presents static gain, as a function of the duty cycle D, for the dc-dc converter cregarded in this chapter, for operation in continuous conduction mode (CCM). Applying the results from Table 2 in (12), it is possible to describe the effective inclination angle θ, for each converter, as a variable dependent on the duty cycle D, as consequence, Table 3 is obtained.

Power dc-dc converter	**Static Gain**
Buck	$G=D$
Boost	$G=\frac{1}{1-D}$
Buck-boost, Cuk, SEPIC and zeta	$G=\frac{D}{1-D}$

Table 2. Static gain for some dc-dc converters in CCM.

Power dc-dc converter	**Effective load inclination angle θ**
Buck	$\theta=atan\left(\frac{D^2}{R}\right)$
Boost	$\theta=\text{atan}\left[\frac{1}{(1-D)^2R}\right]$
Buck-boost, Cuk, SEPIC and zeta	$\theta=\text{atan}\left[\frac{D^2}{(1-D)^2R}\right]$

Table 3. Load curve inclination angle as a function of the converter duty cycle D.

Theoretically, since the duty cycle is limited between 0 and 1, the effective inclination angle becomes restricted into a range whose extremes are dependent on the considered dc-dc converter. For instance, when buck converter is taken into account, for null duty cycle, D=0, (13) is found.

$$\theta\mid_{D=0}=\text{atan}\left(\frac{0^2}{R}\right)=0 \tag{13}$$

Thereby, if the duty cycle is set on its high value, D=1, (14) may be written.

$$\theta\mid_{D=1}=\text{atan}\left(\frac{1}{R}\right) \tag{14}$$

In other to extend the presented analysis for further converters, a similar procedure may be applied, resulting at Table 4 and Figure 14.

From Table 4 it is noticed that effective load inclination angle defines an area on the I-V plan where the maximum power can be tracked. For a better understanding, Table 4 is graphically explained through Figure 14, from where two distinct regions are identified: tracking and non tracking regions.

The tracking region refers to the area on the I-V plan in which the dc-dc converter is able to emulate a proper effective load curve in order to intercept the I-V curve exactly on the MPP, ensuring the maximum power transfer. Note, when solar radiation or temperature change, the maximum power point is relocated on the I-V plan, thus, the effective load inclination angle must also be modified in order to reestablish the maximum power transfer. However, this condition is only suitable if the MPP is located inside the tracking region, otherwise, the system operating point will be set out of the MPP.

Power dc-dc converter	Minimum effective load inclination angle	Maximum effective load inclination angle
Buck	$\theta\|_{D=0}=0$	$\theta\|_{D=1}=atan\left(\frac{1}{R}\right)$
Boost	$\theta\|_{D=0}=atan\left(\frac{1}{R}\right)$	$\theta\|_{D=1}=90°$
Buck-boost, Cuk, SEPIC and zeta	$\theta\|_{D=0}=0$	$\theta\|_{D=1}=90°$

Table 4. Minimum and maximum effective load inclination for some dc-dc converters.

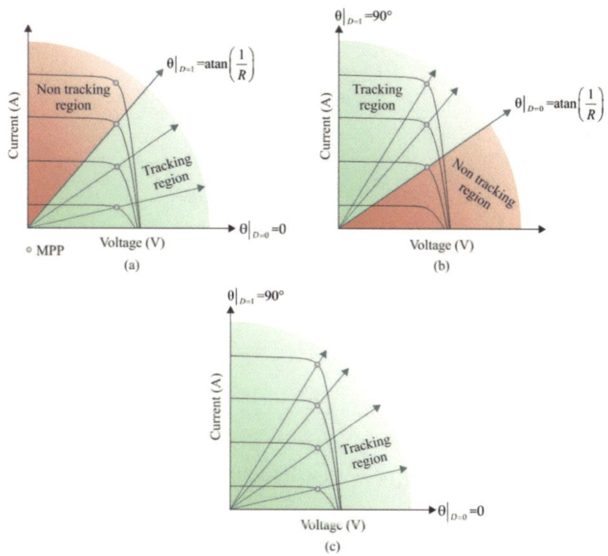

Figure 14. Tracking and non tracking regions for: (a) buck converter; (b) boost converter and (c) buck-boost, Cuk, SEP-IC and zeta converters.

Comparing the graphical results, it is verified that buck-boost, Cuk, SEPIC and zeta are the most appropriated converters for maximum power point tracking applications, once they may track the MPP independently on its position on the I-V plan. On the other hand, buck and boost converters are not indicate for this proposal, since their tracking area is only a part of the whole I-V plan. In order to validate the proposed theory, buck, boost and buck-boost converters were designed and assembled in laboratory, according to Figure 15.

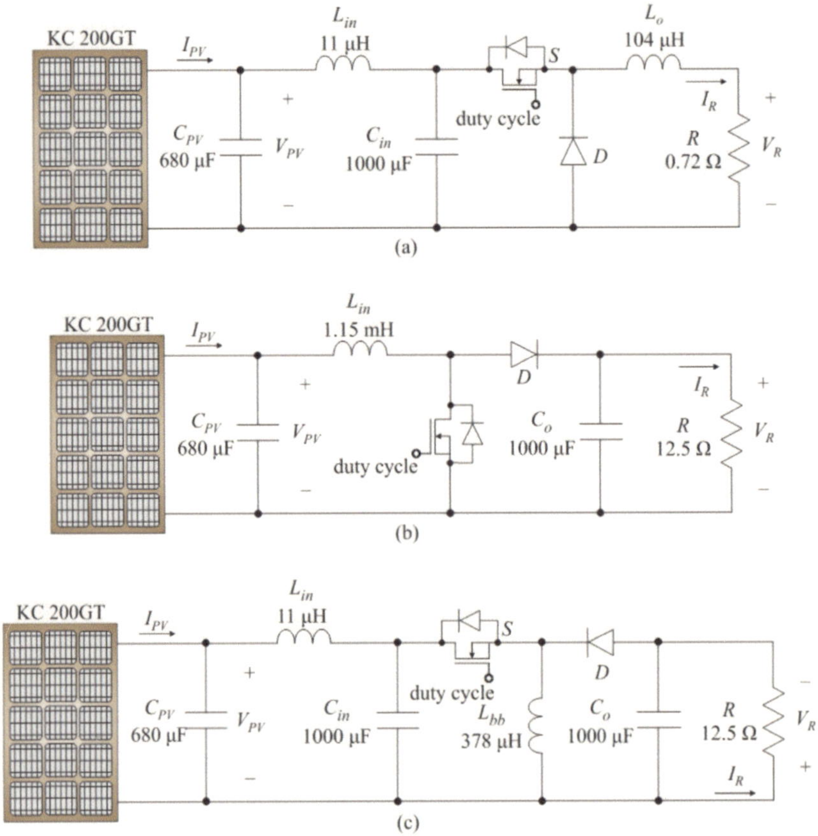

Figure 15. a) buck, (b) boost and (c) buck-boost power stage converters.

For achieving the experimental tests, the converters duty cycle was linearly varied from 0 to 1, while PV voltage and current were measured. By the use of a scope on XY mode, the I-V curve was traced, and the found results are shown at Figure 16.

Notice that I-V curve is partially plotted on the I-V plan when buck and boost converters are regarded, and on the whole I-V plan, when buck-boost converter is considered. Additionally, the area in which the I-V curves were traced is in accordance with the tracking region, theoretically defined for each converter, validating the analysis.

On the next subsection, the resistive load will be replaced by a constant voltage load-type. This analysis is relevant and mandatory, since in many applications, PV systems are employed in battery charges, or even, delivering power to a regulated dc bus.

3.1.2. *Analysis for constant voltage load-type*

The analysis concerning to dc-dc converters operating as MPPT when a constant voltage load-type is considered follows the same procedures presented for resistive loads. For beginning, consider the MPPT system shown in Figure 17, in which a dc voltage source is supplied by a PV module through a literal dc-dc converter.

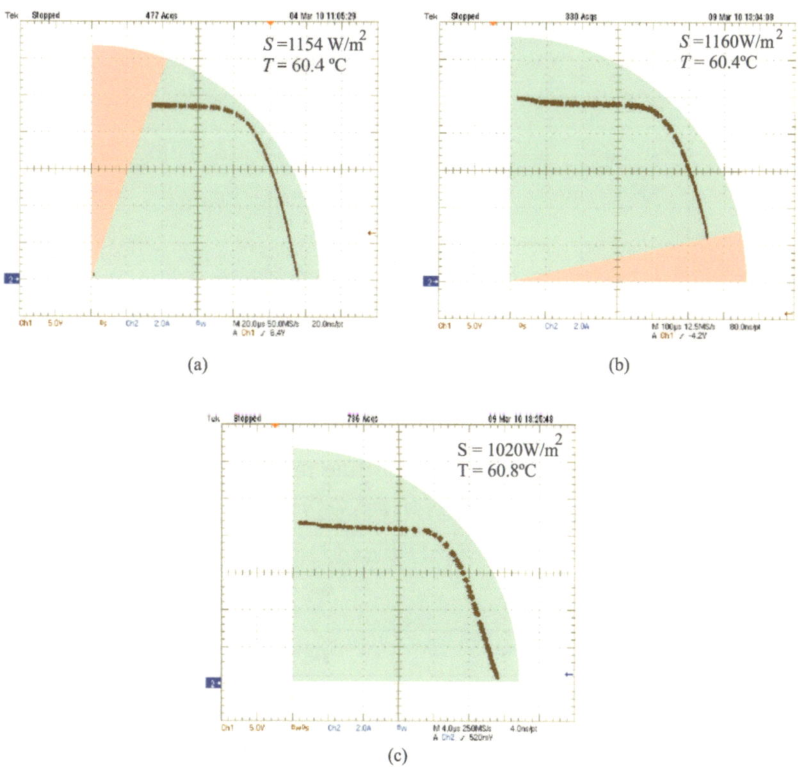

Figure 16. Experimental results for (a) buck, (b) boost and (c) buck-boost converters.

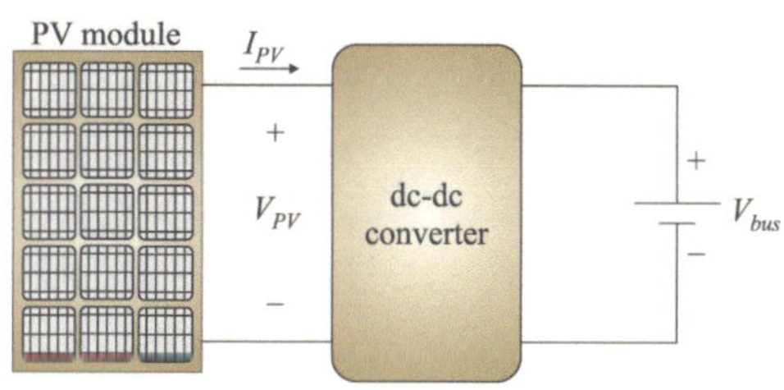

Figure 17. MPPT supplying a constant voltage load-type.

In this case, the output converter voltage is imposed by the load, permitting to write (15) and to model both, dc-dc converter and voltage load, as a controlled voltage source, as is shown in Figure 18.

$$V_{PV} = \frac{V_{bus}}{G} \tag{15}$$

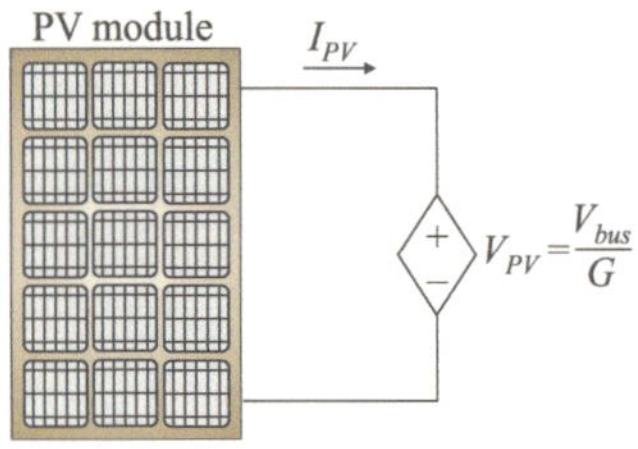

Figure 18. Equivalent MPPT system obtained from the PV module terminals.

Taking into account the static gain *G* presented at Table 2, it is possible to define the equivalent voltage source for each dc-dc converter as a function of the duty cycle *D*, resulting on Table 5. Due to the duty cycle range restriction, $0<D<1$, the voltage imposed by the equivalent controlled voltage source across the PV module terminals is also limited. For example, when buck converter is regarded, equations (16) and (17) are obtained from the first line of Table 5, describing the system behavior for $D=0$ and $D=1$, respectively.

Power dc-dc converter	Equivalent voltage source value
Buck	$V_{PV} = \frac{V_{bus}}{D}$
Boost	$V_{PV} = (1-D)V_{bus}$
Buck-boost, Cuk, SEPIC and Zeta	$V_{PV} = \frac{(1-D)}{D} V_{bus}$

Table 5. Minimum and maximum effective load inclination for some dc-dc converters.

$$V_{PV} \big|_{D=0} = \frac{V_{bus}}{0} \rightarrow \infty \tag{16}$$

$$V_{PV} \big|_{D=1} = V_{bus} \tag{17}$$

It is important to emphasize that the maximum voltage across the PV module terminals is its open circuit voltage, thus, the minimum duty cycle value must be defined in order to satisfy this condition. From the exposed, (16) is replaced by (18).

$$V_{PV}\Big|_{D=D_{\min}} = \frac{V_{bus}}{D_{\min}} = V_{oc} \tag{18}$$

Extending the analysis for further converters, Table 6 is obtained.

The graphical representation allows understanding how the dc-dc converter feature impacts on the tracking quality when constant voltage loads are employed, as depicts Figure 19. When the maximum power point is located inside the tracking region, the dc-dc converter may apply on the PV output terminals a voltage value for ensuring its operation on the MPP. Otherwise, even when the better tracking algorithm is used, there is no possible to track it.

Power dc-dc converter	Minimum voltage across the PV module terminals	Maximum voltage across the PV module terminals
Buck	$V_{PV}\big\|_{D=1}=V_{bus}$	$V_{PV}\big\|_{D=D_{min}}=V_{oc}$
Boost	$V_{PV}\big\|_{D=1}=0\ \text{V}$	$V_{PV}\big\|_{D=0}=V_{bus}$
Buck-boost, Cuk, SEPIC and zeta	$V_{PV}\big\|_{D=1}=0\ \text{V}$	$V_{PV}\big\|_{D=D_{min}}=V_{oc}$

Table 6. Minimum and maximum voltage values across the PV module terminals.

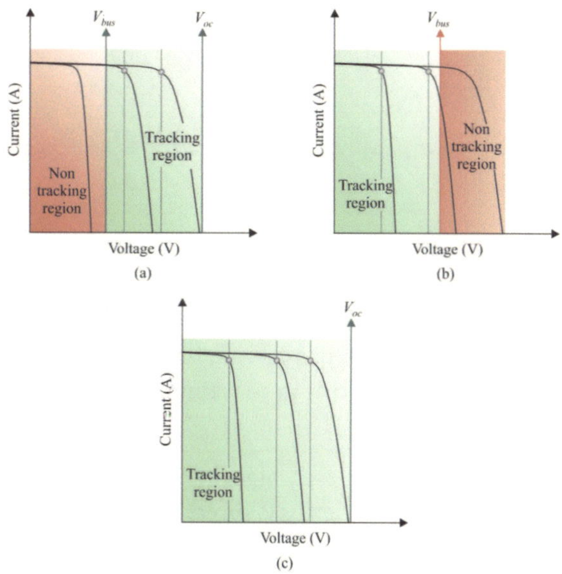

Figure 19. Tracking and non tracking regions for: (a) buck converter; (b) boost converter and (c) buck-boost, Cuk, SEPIC and zeta converters.

In addition, notice that temperature changes may directly affect the tracking quality: commonly, PV systems are designed considering its operation on the SCT, i.e., $T = 25^\circ$, however, when PV modules are exposed to the solar radiation, its real temperature of operation encreases and, as consequence, the voltage associated to MPP is moved to the left. This behavior is critical for buck converter, as per Figure 19 (a), since its non tracking region is also on left of V_{bus}.

Although boost converter also presents a non tracking region, in this case it presents a proper tracking behavior, once temperature increasing replaces the MPP to left, toward to the tracking region, in according to Figure 19 (b).

Finally, buck-boost converter (and similars) can track the MPP independent on its position on the I-V pan, as is shown on Figure 19 (c). Furthemore, these converters are also indicated for tracking applications, when constant voltage loads are employed.

3.2. MPPT from the tracking algorithm point of view

The tracking algorithm performance is fundamental for an efficient tracking response. Usually, the algorithm receives the PV module voltage and current as input data and defines the dc-dc converter duty cycle that establishes the system operating point on the MPP, as depicts Figure 20.

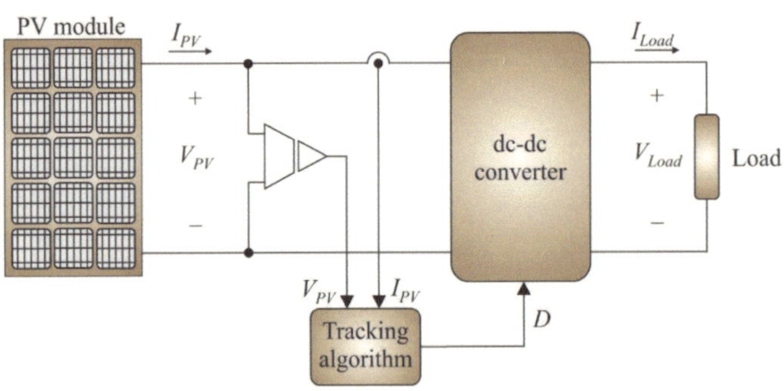

Figure 20. Typical input and output data related to MPPT algorithms.

As the radiation and temperature are dynamic variables, and the MPP depends on both of them, the algorithm must practically work in real time, updating the duty cycle for a fast and accurate tracking.

On the literature, there are several proposed algorithms for improving the tracking speed, accuracy or both, but the algorithm efficiency is directly associated to the complexity of implementation.

In this section, based on the PV curves understanding, a new tracking method is developed, whose main characteristics are: simplicity, excellent tracking dynamic, accuracy, stability in steady-state (no oscillations), and low cost.

Before presenting this proposal, a review of the most commonly employed MPPT algorithms is presented, where Constant Voltage, Perturb and Observe (P&O) and Incremental Conductance (IncCond are briefly discussed.

3.2.1. Constant Voltage

This method is achieved in or to impose the voltage across the PV terminals clamped at a fixed value, normally specified to ensure the maximum power transfer on the STC [4]. Once a single voltage sensor is needed, it is featured by simplicy of implementation and low cost, but for any temperature change, the PV operating point is set out of the MPP.

3.2.2. Perturb and Observe

Perturb and Observe (P&O) is one of the most diffused MPPT algorithms, whose tracking response is independent on the environmental conditions, however, its implementation requires a voltage and a current sensor, increasing the cost and complexity [11].

When in operation, the P&O algorithm calculates the PV output power and perturbs the converter duty cycle (increasing or decreasing it). After perturbation, the PV output power is recalculated and, if it was increased, the perturbation is repeated on the same direction, otherwise, it is inverted.

The main drawbacks associated to this method are the oscillation in steady-state, due to the constant perturbations, the slow tracking dynamic and the inability to proper operate during fast changes of solar radiation.

3.2.3. Incremental Conductance

The Incremental Conductance (IncCond) method is featured for combining both, tracking speed and accuracy [13]. From the voltage $V_{PV}$and current I_{PV} measurements, the algorithm calculates the photovoltaic output power P_{PV} and its derivative in function of the voltage dP_{PV}/dV_{PV}, using both results to define if the duty cycle must be increased or decreased, in order to impose the system operating point on the MPP. Usually the IncCond method is implemented digitally, and the derivative is calculated by the microcontroller according to (19).

$$\frac{dP_{PV}}{dV_{PV}} = I_{PV}(n) + V_{PV}(n)\frac{I_{PV}(n-1)-I_{PV}(n)}{V_{PV}(n-1)-V_{PV}(n)} \tag{19}$$

From (19), the following decision logic is achieved:

a. if $dP_{PV}/dV_{PV}0$ (left of MPP), the duty cycle is changed for elevating the PV module voltage;

b. if $dP_{PV}/dV_{PV}0$ (right of MPP), the duty cycle is alterated for decreasing the output voltage;

c. if $dP_{PV} / dV_{PV} = 0$ (at MPP), the duty cycle is maintened unchangeable.

This tecnique is characterized for presenting high tracking speed (variabe step) and accuracy (no oscillation), however, it is more complexy than P&O, once the derivative must be calculated in real time and also require a voltage and a current sensor.

3.2.4. A MPPT algorithm based on temperature measurement

The MPPT algorithm based on temperature measurement, named by MPPT-temp [22], consists on the unification of simplicity related to Constant Voltage method with the velocity and accuracy tracking related to the Incremental Conductance technique.

The development of this method comes from (3), rewritten in (20). Note that the voltage in which the maximum power is established depends exclusively on the PV surface temperature. Thereby, accomplishing the temperature measurement, the maximum power voltage V_{mpp} may be determined and actively imposed across the PV terminals in real time. The configuration needed for implementation of this new method is depicted at Figure 21.

$$V_{mpp} = V_{mpp}^{STC} + \left(T - T^{STC}\right)\mu_V \tag{20}$$

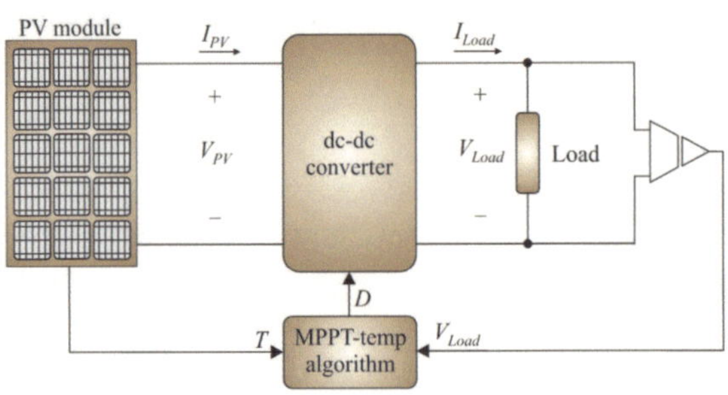

Figure 21. Configuration of the new tracking method.

The following steps describe how the maximum power point is achieved by the proposed method:

1. The PV module surface temperature T and the load volatge V_{Load} are measured by a temperature sensor and a volatge sensor, respectivelly;

2. Both sinals are applyed as input data for the tracking algorithm, whose output is the dc-dc converter static gain G_{mpp}, in acordance with (21);

$$G_{mpp} = \frac{V_{Load}}{V_{mpp}^{STC} + \left(T - T^{STC}\right)\mu_V} \tag{21}$$

3.Defining the dc-dc converter, an expression for determine the duty cycle may be derived. For example, considering a buck converter, i.e. $G=D$, the maximum power point is acomplished when (21) is rewritten as (22).

$$D_{mpp}=\frac{V_{Load}}{V_{mpp}^{STC}+\left(T-T^{STC}\right)\mu_{V}} \tag{22}$$

For extending the analisys, Table 7 present the equitions employed by the tracking algorithm to calculate the duty cycle that imposes the PV operating point on the MPP.

Power dc-dc converter	Duty cycle for operation on the MPP
Buck	$D_{mpp}=\frac{V_{Load}}{V_{mpp}^{STC}+\left(T-T^{STC}\right)\mu_{V}}$
Boost	$D_{mpp}=1-\frac{V_{Load}}{V_{mpp}^{STC}+\left(T-T^{STC}\right)\mu_{V}}$
Buck-boost, Cuk, SEPIC and zeta	$D_{mpp}=\frac{V_{Load}}{V_{Load}+V_{mpp}^{STC}+\left(T-T^{STC}\right)\mu_{V}}$

Table 7. Duty cycle for maximum power point operation.

Table 7 evidences the simplicity of the proposed method: from the temperature measurement it is possible directly set the duty cycle, ensuring the maximum power transfer from the PV module to the load. Additionally, there is no need computational requirements, recursive procedures and, due to the slow dynamic associated to the temperature changes, the tracking is smooth, stable and occurs in real time, for any combination of solar radiation and temperature.

It is important to emphasize the restrictions associated to the dc-dc converter, as it was previously studied at last section, may imply in a poor tracking quality, even when the MPPT-temp method is employed. In order to finalize the chapter, the results obtained from an experimental prototype are presented.

The prototype was implemented for processing the power generated by a KC200GT PV module, from Kyocera, whose electrical specifications are listed at Table 1. A dc-dc buck-boost converter was chosen for composing the hardware, because of its proper tracking behavior. Consequently, the equation to define the duty cycle is given by (23).

$$D_{mpp}=\frac{V_{Load}}{V_{Load}+V_{mpp}^{STC}+\left(T-T^{STC}\right)\mu_{V}}=\frac{V_{Load}}{V_{Load}+29.8-0.14T} \tag{23}$$

For measuring the PV module surface temperature, a precision centigrade temperature sensor (LM35) was used. This device presents a linear gain of 10 mV/°C. Furthermore, in order to execute the tracking algorithm, a simple PIC 18F1320 microcontroller was employed.

The tracking method validation was achieved plotting the PV operating point on the I-V plan during temperature and solar radiation changes, as is illustrated in Figure 22. For reaching this result a scope on XY mode, where X refers to the PV output voltage and Y refers to the PV output current, was employed.

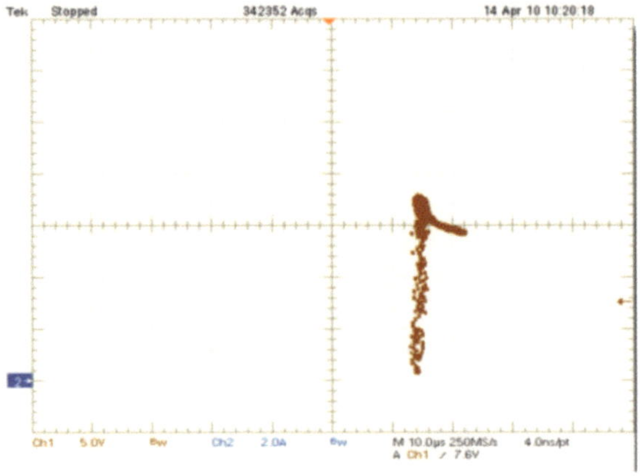

Figure 22. Experimental PV operating point on I-V plan during temperature and solar radiation changes. Scope on XY mode (X-voltage, Y-current)

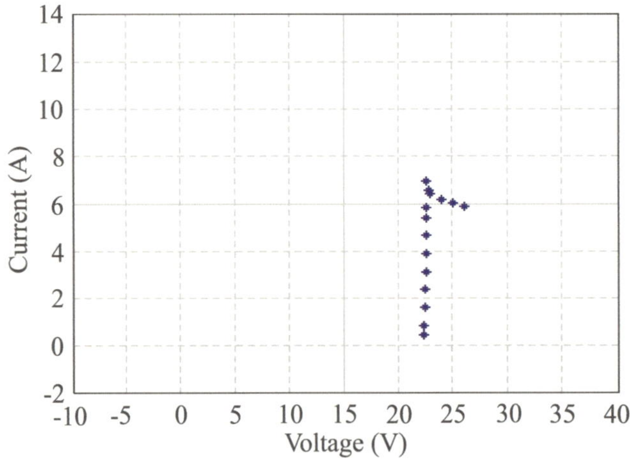

Figure 23. Theoretical MPP trajectory for solar radiation and temperature measured during experimental evaluation.

In order to prove that the obtained experimental trajectory coincides with the MPP, the values of solar radiation and temperature were also collected by a data logger and summarized by Table 8. The measurements were achieved on middle of April in the Florianopolis Island – south of Brazil – located at latitude 27°.

Solar radiation	Temperature
900 W/m^2	51°C
850W/m^2	50°C
830W/m^2	49°C
802W/m^2	41°C
787W/m^2	34°C
770W/m^2	26°C
700W/m^2	51°C
600W/m^2	51°C
500W/m^2	51°C
400W/m^2	51°C
300W/m^2	52°C
200W/m^2	52°C
100W/m^2	53°C
50W/m^2	53°C

Table 8. Solar radiation and temperature values obtained from data logger during experimental tests.

From Table 8, and employing (3) and (4), the theoretical voltage and current values for the system operating on the MPP were determined and plotted on the I-V plan, resulting at Figure 23, from where it is possible to conclude that the trajectory obtained by experimentation is equivalent to the trajectory described by the maximum power point.

4. Conclusion

Currently the most common employed PV tracking algorithm are Perturb and Observe and Incremental Conductance. Perturb and observe is simple, however it failures to track the MPP under abrupt changes on solar radiation and presents oscillations around the MPP on steady-state. Incremental conductance is accurate, however, it implementation is more complex. In these both algorithms, it is necessary to measure the PV output voltage and current.

The proposed algorithm is simpler than perturb and observe and more accurate than the incremental conductance. Furthermore, the current sensor is substituted by a temperature sensor, implying in cost reduction. Eliminating the current sensor means avoid the output power high frequency variation associated to the PV current (directly proportional to solar radiation), thus, there is no oscillation around the MPP. Since the tracking is based on the temperature, its low dynamic ensures a soft tracking, but accurate and fast.

Based on the exposed, this tracking method employment allows a significant improvement on the PV operation. Note that in PV generation the power losses may be separated in two terms: the first one is associated to the power converters efficiency and the second to the tracking algorithm efficiency. Thus, if an optimized tracking algorithm is employed, the global efficiency is increased.

From the dc-dc analysis it was verified that buck and boost converters, even been largely applied for tracking applications, are not proper for this proposal, since they can track the MPP only in a part of the I-V plan. On the other hand, buck-boost converter (and Cuk, SEP-IC or zeta) may track the MPP around the whole I-V plan, configuring the better option for tracking applications.

Finally, an optimized tracking system only is obtained if both, dc-dc converter and tracking algorithm properly operate. This condition is accomplished when a buck-boost converter is employed in combination with the MPP-temp tracking method.

Acknowledgements

The authors would like to thank CNPq and FINEP for financial support, and Power Electronics Institute, for technical support. Additionally, the authors would like to thank the Eng. Walbermark M. dos Santos, for his several contributions.

Author details

Roberto Francisco Coelho* and Denizar Cruz Martins

*Address all correspondence to: roberto@inep.ufsc.br

Federal University of Santa Catarina - Electrical Engineer DepartmentPower Electronics Institute, Florianópolis, Brazil

References

[1] Jung-Min, K., Kwang-Hee, N., & Bong-Hwan, K. (2006). Photovoltaic Power Conditioning System With Line Connection.IEEE. *Transactions on Industrial Electronics*, 53, 1048-1054.

[2] De Souza, K. C. A., dos Santos, W. M., & Martins, D. C. (2010). Active and reactive power control for a single-phase grid-connected PV system with optimization of the ferrite core volume. *IEEE/IAS International Conference on Industry Applications (INDUSCON)*, 1-6.

[3] Ciobotaru, M., Teodorescu, R., & Blaabjerg, F. Control of single-stage single-phase PV inverter. *European Conference on Power Electronics and Applications*, 10.

[4] Ghaisari, J., Habibi, M., & Bakhsahi, A. (2007). An MPPT Controller Design for Photovoltaic (PV) System Based on the Optimal Voltage Factor Tracking. *IEEE Canada electrical Power Conference*, 359-362.

[5] Pandey, A., Dasgupta, N., & Mukerjee, A. K. (2007). A Single-Sensor MPPT Solution. *IEEE Transaction on Power Eletronics*, 22(2), 698-700.

[6] Sokolov, M., & Shmilovitz, D. (2008). A modified MPPT Scheme for Accelerate Convergence. *IEEE Transactions on Energy Conversion*, 23(4), 1105-1107.

[7] Villalva, M. G., & Ruppert, E. F. (2009). Analysis and Simulation of the P&O MPPT Algorithm Using a Linearized PV Array Model. *10th Brazilian Power Electronics Conference*, 231-236.

[8] Aranda, E. D., Galán, J. A. G., Cardona, C. S., & Marques, J. M. A. (2010). Measuring the I-V Curve of PV Generator. *IEE Industrial Electronics Magazine*, 3(3), 4-14.

[9] Coelho, R. F., Concer, F. M., & Martins, D. (2009). *A Study of the Basic DC-DC converters Applied in Maximum Power Point Tracking. 10th Brazilian Power Electronics Conference*, 673-667.

[10] Hohm, D. P., & Ropp, M. E. (2000). Comparative Study of Maximum Power Point Tracking Algorithms Using an Experimental, Programmable, Maximum Power Point Test Bed. *IEEE Photovoltaic Specialists Conference*, 1699-1702.

[11] Tan, C. W., Green, T. C., & Hernandez-Aramburo, C. A. (2008). Analysis of Perturb and Observe Maximum Power Point Tracker Algorithm for Photovoltaic Applications. *IEEE 2nd International Power and Energy Conference*, 237-242.

[12] Boico, F., & Lahman, B. (2006). Study of Different Implementation Approaches for a Maximum power Point Tracker. *IEEE Computers in Power Electronics*, 15-21.

[13] Liu, B., Duan, S., Liu, F., & Xu, P. (2007). Analysis and Improvement of a Maximum Power Point Tracking Algorithm Base on Incremental Conductance Method for Photovoltaic Array. *IEEE International Conference on Power Applications*, 637-641.

[14] Yuvarajan, S., & Shoeb, J. (2008). A Fast and Accurate Maximum Power Point Tracker for PV Systems. *IEEE Applied Power Electronics Conference and Exposition*, 167-172.

[15] Femia, N., Petrone, G., Spagnuolo, G., & Vitelli, M. (2004). Optimizing Sampling Rate of P&O MPPT Technique. *IEEE Power Electronics Specialist Conference,*, 3, 1945-1949.

[16] Pandey, A., Dasgupta, N., & Mukerjee, A. (2006). Designe Issues in Implementing MPPT for Improved Tracking and Dynamic Performance. *IEEE Conference on Industrial Electronics*, 4387-4391

[17] Onat, Nevzat. (2010). Rececent developments in maximum power point tracking technologies for photovoltaic systems. *International Journal of Photoenergy*.

[18] De Brito, M. A. G., Junior, L. G., Sampaio, L. P., Melo, G. A., & Canesin, C. A. (2011). Main maximum power point tracking strategies intended for photovoltaics. *XI Power Electronics Brazilian Conference*, 524-530.

[19] Integration of alternative sources of Energy. (2006). Felix A. Ferret and M. Godoy Simões. Wiley-Intercience. *IEEE Press, New Jersey*.

[20] http://www.physicalgeography.net/fundamentals/7f.html.

[21] Coelho, R. F., Concer, F. M., & Martins, D. (2009). A Proposed Photovoltaic Module and Array Mathematical Modelling Destined to Simulation. *IEEE International Symposium on Industrial Electronics*, 1624-1629.

[22] Coelho, R. F., Concer, F. M., & Martins, D. C. (2010). A MPPT approach based on temperature measurements applied in PV systems. *IEEE/IAS International Conference on Industry Applications*, 1-6.

Chapter 5

Industrial Energy Auditing for Increased Sustainability – Methodology and Measurements

Jakob Rosenqvist, Patrik Thollander,
Patrik Rohdin and Mats Söderström

Additional information is available at the end of the chapter

http://dx.doi.org/10.5772/51717

1. Introduction

Industrial energy efficiency is a key component in the transition of the economy towards increased sustainability. For an industrial company, there are four means to reduce energy costs: implementing energy-efficient technologies, energy carrier conversion, load management, and more energy-efficient behaviour. The European end-use energy efficiency and energy services directive promotes, among other things, the removal of existing market barriers and imperfections that impede the efficient end use of energy [1]. Energy audits provide an important tool in reducing barriers to energy efficiency [2]. Furthermore, an initial, well-structured energy audit is the first important step in a successful in-house energy management program in industry [3].

From the global perspective, industrial energy efficiency is one of the most important means of reducing the threat of increased global warming [4] as the industry accounts for about 80 percent of the world's annual coal consumption, 40 percent of the world's electricity use, 35 percent of the world's natural gas consumption, and around 10 percent of global oil consumption [5]. Of great importance are thus different means which promote energy efficiency for the industrial sector. In Europe, growing concern for increased global warming has led to the implementation of a number of policy instruments such as the EU Emission Trading Scheme (ETS) and the European Energy End-Use Efficiency and Energy Services Directive (ESD). From the industry's perspective, supply side policy instruments like the EU ETS will most likely result in higher European energy prices which will stress the industry to take actions toward increased energy efficiency. On the other hand, this may lead to competitive disadvantages compared to industries outside the EU [6].

The role and importance of energy audits varies from country to country. In a comparing study of factors influencing energy efficiency in the German and Colombian manufacturing industries [7], firms and associations were asked about their view on energy audits. Among the German respondents 57 percent states that volontary audits is an important factor influencing the energy efficency in their country, while 61 percent of the Colmbian respondents states that volontary audits is an important or very important factor. When they were asked about their own measures and actions, 71 percent of the German respondents and 54 percent of the Colombian respondents stated that they would consider energy audits to increase energy efficiency performance.

This audit method was developed in Sweden, in the Swedish manufacturing industry context. Evaluations of conducted energy audits in small and medium-sized manufacturing companies shows that the calculated technical potential for increasing energy efficiency performance varies from 16-40% of the total energy use. For electricity the calculated potential is up to 60% [8]. If the suggested measures are implemented and if the potential is reached or not depends on the barriers and driving forces for energy efficiency. The energy price can play an important role. For Swedish industry, energy prices have risen significantly in recent years. Between 2000 and 2006 electricity prices in Swedish industry almost doubled and oil prices rose by about 70 percent [9,10]. This trend has not declined. In January, 2010, prices on the Nordic electricity spot market arose to some 130 Euro per MWh. The electricity price increases were partly due to the deregulation of the European electricity markets as the deregulation has caused the national markets to converge and Sweden has for a long time enjoyed one of the lowest electricity prices in Europe [11]. While the oil price increases may not create competitive disadvantages for Swedish industry, the electricity price increases most likely will, as this is particularly related to the Swedish industry and the fact that the previously low electricity prices have resulted in a higher use of electricity in many Swedish industrial sectors compared to their European competitors.

The methodology described is primarily for a technical energy audit with the aim to make the energy use more efficient and sustainable. In the audit, organizational issues and the surrounding society are regarded mainly as means to reach the goal, and are important driving forces to implement changes.

1.1. Classification of energy audits

Energy audit models can be described in terms of the scope, thoroughness and aim of the audit [12]. The method presented in this chapter normally is used with a broad scope, covering the entire studied site. The thoroughness can vary from rough comb to fine comb and often the thoroughness varies between different unit processes in the same audit. As a top-down approach is used, the auditor always starts with a broad scope and a rough comb. As the audit progresses and the key areas are pointed out, the data collection and analysis become more detailed. Lytras [12] also describes the aim of an energy audit as either to point out general energy saving areas or to propose specific energy saving measures. With the method described in this chapter, pointing out the general energy saving areas is one way to identify specific measures and prioritize them.

http://dx.doi.org/10.5772/51717

Energy audits can also be described with the classification of energy audits from ASHRAE [13]. This classification involves three levels, defined in reference [14], and is similar to the audit procedure presented in reference [15]. The three levels presented in [13] are Level I (Walk-through assessment), Level II (Energy survey and analysis), and Level III (Detailed analysis of capital-intensive modifications).

The Level I walk-through assessment involves an assessment of the energy cost and efficiency by analyzing energy bills and a brief survey of the site. This first-level assessment targets low- or no-cost measures and presents a listing of capital improvements that need to be studied further. Level II, energy survey and analysis, includes a more detailed survey and analysis of the plant studied [13]. This is usually done by some form of detailed breakdown of energy use, either in activities and energy carriers or, as in this chapter, in unit processes. An energy audit with the method described in this chapter always involves the assessment at Level I and Level II.

1.2. Unit processes

A unit process is defined by the energy service to be performed and industrial processes may thus be divided into two categories of unit processes:

- Production processes – the processes needed to manufacture the products.
- Support processes – the processes needed to support the production processes but not needed for production.

Production processes	Disintegrating	Support processes	Administration
	Disjointing		Cooling
	Mixing		Lighting
	Jointing		Compressed Air
	Coating		Ventilation
	Moulding		Pumping
	Heating		Tap water heating
	Melting		Internal Transport
	Drying		Space Heating
	Cooling/Freezing		Steam
	Packing		

Table 1. The unit processes used at Linköping University, divided into production and support processes.

Unit processes are a way to divide the energy use of an industry or other businesses into smaller parts. The unit process perspective also enables one to question the methods used for different processes. Air flows in ventilation systems are sometimes high because the ventilation system may be dimensioned both for ventilation and cooling. If you see that the device is taking care of two different processes you also see that there is a point in adjusting the flow depending on the cooling demand.

The concept is based on the objective of the industrial process, the mixing of materials, cooling or drying a product, the production of compressed air or to carry goods, etc. Unit processes are considered to be the smallest components within an industrial energy system. Unit processes are general for all industries, thus providing opportunities for comparisons of a given unit process, such as forming, between different industries. Using the unit processes to represent the "building blocks" of energy use also enables modelling, e.g. simulation or optimization modelling of industrial energy use. For example, see [16].

2. Methodology

This section describes one method to survey and analyze the energy use of an industry. The industrial energy audit may be divided into three main parts: the survey, the analysis and the proposed measures. The expected result of the energy audit is a number of measures that will increase the energy efficiency, switch from non-renewable to renewable sources and decrease the energy use of the company or in the energy system as a whole. The result is normally presented in a report.

In the three parts of the audit you have different tasks:

Energy survey – What do we have?

- Define the system boundary.
- Identify unit processes.
- Quantify energy supply.
- Allocate energy use to different unit processes.

Energy analysis – How "bad" is the utilization of the supplied energy? Where does it go wrong? How far can we reach?

- Identify both system errors and detail problems.
- Identify idling power and energy.
- Identify technologies that are far from BAT (Best Available Technology).
- Identify mismatching energy quality, where high quality energy is used for non-demanding purposes, for example, electricity for low temperature heating.
- Identifying barriers and driving forces.
- Sum up the potential for energy efficiency and conversion for the system as a whole.

Suggested measures – What can we do about it?

- Identify possible solutions.
- Calculate the impact of the solutions and analyze them.
- Calculate the economic impact.

2.1. Top-down approach and iterative method

The energy audit is basically a project with the aim to suggest energy efficiency and conversion measures. The basic method is the project method, but the actual workflow does not follow the straight project line from survey to measures. In practice you will start analyzing problems and identifying solutions as soon as you start collecting information for the survey. The analysis and ideas for measures will also affect the need for data collection. Ideally, the method may be described as an iterative process, as shown in Figure 1, where the iterations stop when you have enough data to suggest relevant measures.

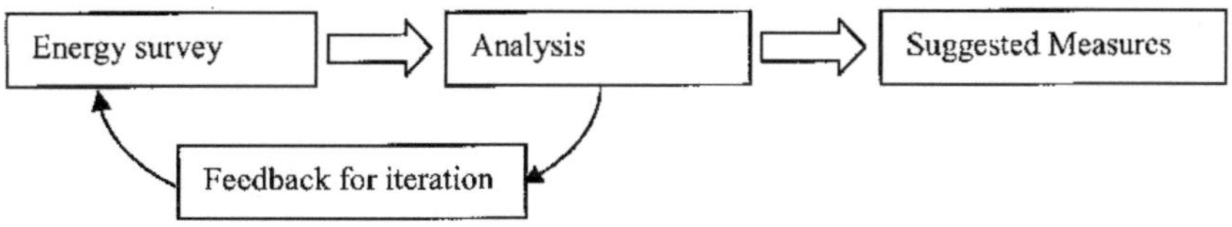

Figure 1. Schematic workflow for an industrial energy audit.

The iterative method is combined with a top-down approach. In this case, it means that the studied object (organisation, business or site) is initially considered as a unit, with its streams of energy and materials (Figure 2). The total use of different kinds of purchased energy is quantified. The material use and the products may be described but not necessarily quantified.

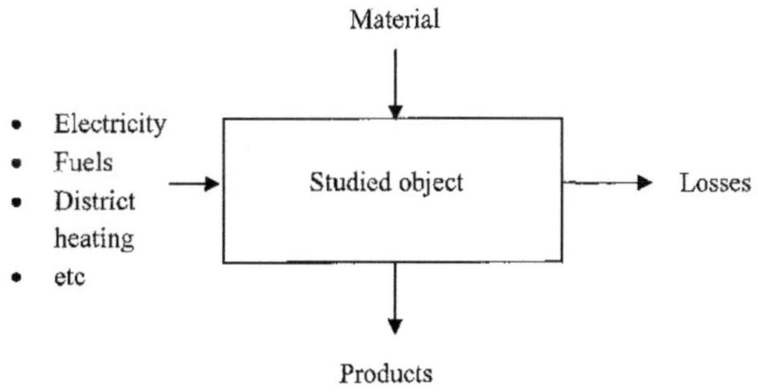

Figure 2. The top level of the top-down perspective for an energy audit.

The next step is to describe the unit processes and analyse them in terms of energy use (Figure 3). The first overview of the energy use makes it possible to relatively quickly identify key areas where additional resources are needed for measurements and analysis [17].

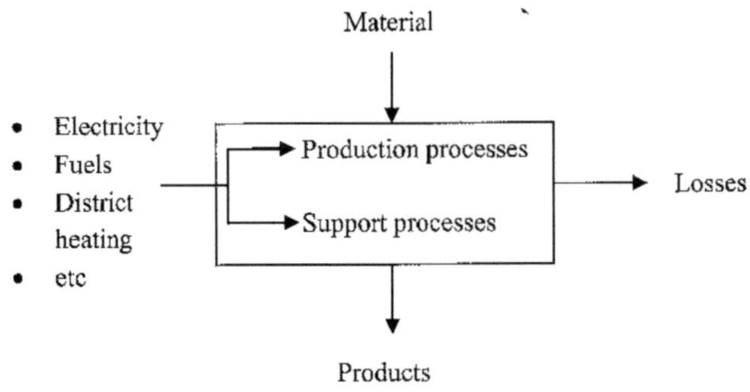

Figure 3. The second level of the top-down perspective for an energy audit means going into more detail concerning unit processes.

The next possible level is to find the data that enables you to allocate the energy use to the relevant unit processes. Which processes are relevant depends on the purpose of the audit. The top-down approach can continue further down, to individual components if necessary. How much detail the investigation will involve depends on the purpose of the audit, the possibility to acquire data, the complexity of the suggested measures, the timeframe for the audit, etc.

The point of starting from the top and working your way down is that the detailed information should be provided where the auditor thinks that it is needed, but not necessarily everywhere. The level of detail will vary between different unit processes. Sometimes the energy of all unit processes in a production line is summarized under the heading of Production Processes. It can be done for practical reasons, such as when a machine with only one energy supply handles multiple unit processes, or because the audit is delimited to focus only on support processes.

2.2. Typical process for an audit

In practice the activity schedule for a project can look like this:

Week 1: Start meeting with the company for project start and start collecting information. Wait for data from the company. Describe the unit processes. Collect basic data, like annual energy statistics (from energy supplier) and if available figures from sub-metering at the plant, as well as technical drawings. Finally, analyze the collected material. For example, where may the largest energy efficiency potentials be located? Where can we probably measure what we want?

Week 2: First visit on site – identify energy supply to different unit processes, collect data, measure, put loggers in place, count equipment.

Analyse data from visit. Detailed data is now available, but a top-down perspective is still applied, e.g. where may the largest energy efficiency potentials be located, which unit processes are most promising and in which areas do we need more data or knowledge?

Week 3: Second visit on site – remove loggers, additional instantaneous measurements. Analyse data from loggers. Completing the survey.

http://dx.doi.org/10.5772/51717

Week 4: Calculate the energy efficiency potential of the measures. Find suppliers of equipment needed for the measures, and wait for data from them. Complete the report.

Week 5: Third visit on site – present the results. Validate the proposed measures with the company staff. Some optimal (in terms of calculation) energy efficiency measures may be switched to second or even third-best choice after discussion with company staff.

Please note that this is a schematic example of how an energy audit could be carried out. The same method can be applied on different scales. A quick Level I audit at a small facility (i.e., five employees and 500 MWh electricity per year) can be finished in two weeks and less than 30 working hours. It would be fully possible to spend several months on a larger and more complex object (i.e., >5 buildings, many energy supplies, several production lines) which demands a more thorough audit and detailed measures.

2.3. Defining the object and the system boundaries

The object for the audit and its system boundaries have to be defined. As it is an energy audit, you will study energy flows, but the focus may be limited. The focus can be on decreasing a supply, e.g. electricity, fossil fuels or purchased energy, or on a part of the object, e.g. the building envelope or support processes only. The system boundaries can be an organisation or a part of an organisation, a building or a physical area with more than one organisation. The perspective can also be broader, for example, taking management or economic aspects or embedded energy into account.

As the method is iterative, you might have to narrow or widen the system boundary, depending on the limitations and possibilities you find on the way. One example is if the organisation you study owns the building where they work, but you discover that they also have a tenant working in a smaller part of the building. Instead of trying to calculate how much energy the tenant uses, the system border can be redefined to cover the tenant too.

Another example is when there is an existing or possible symbiosis between two neighbouring businesses. In this case the system is growing from one to two companies. A very common case is that the audited business is renting the building they work in. In this case you have the choice to exclude the building from the audit, but you will most probably find more and better efficiency measures by treating the building and the business as one system, with the business owner and the landlord as two participants.

2.4. Energy survey and analysis – Introduction

With the energy survey you will be able to create a power balance and an energy balance with the incoming energy allocated to different unit processes. To be able to suggest energy efficiency and conversion measures you also need to understand how the processes work, at least when it comes to energy use. The aim of the energy survey is not necessarily to create a very detailed map of the energy use, but to find the potential for energy efficiency measures and potential conversion to renewable energy sources.

By starting with an overview of the energy system and using a top-down approach it is possible to find the potential changes at the structural level, and avoid being limited to details. By combining the top-down approach with an iterative method for the data collection, it is possible to go further into details without getting overwhelmed with detailed data that is not really needed.

There are different ways to get the data you need. Provided that the top-down perspective is used, you don't have to collect all possible data about every device. The data collection begins with the most accessible data at the most general level which is considered useful by the auditor (using a top-down approach). The analysis starts with analyzing that initial data. The analysis will probably raise questions, which are used to guide the coming data collection procedures.

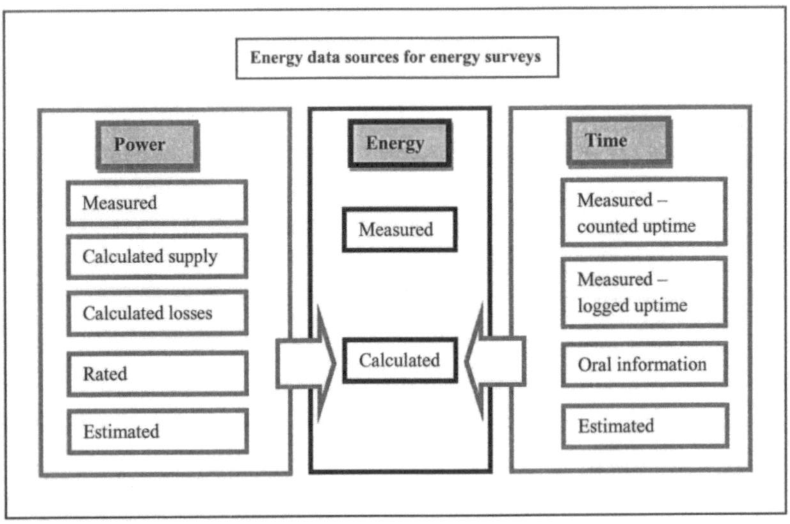

Figure 4. Some possible energy data sources for energy surveys.

At some level you have to stop the data collection despite not having all the answers. The limit is often the time reserved for the project, but it can also be the possibility to measure, practical obstacles to changing a process or the lack of historical data. It is often necessary (and possible) to use data from a more aggregated level, together with some analysis, calculations and assumptions. The data collection procedure could be carried out in the following way:

- Step 1: Gather statistics: electricity, district heating, fuel, water, production rate, etc. Gather drawings and plans.
- Step 2: Visit the company during operating hours.
- Step 3: Visit the company during non-operating hours.
- Step 4: Create a power balance, energy balance and allocate energy use.

To complete the audit you need to take two more steps:

http://dx.doi.org/10.5772/51717

- Step 5: Create an energy balance after adjustments (showing the impact of the suggested measures).
- Step 6: Confirmation of audit results.

The main purpose of the data collection is to give you basic data to identify relevant unit processes, to create a power balance and energy balance for these processes. Another purpose is to get enough data to be able to suggest adequate measures for energy efficiency and conversion. The origin and quality of information can vary greatly. Focusing solely on the energy data, the potential data sources can be categorized as shown in Figure 4.

2.5. Step 1: Gather statistics, drawing and plans

Before you have even visited the object of your survey, you can review the statistics concerning the use of electricity, district heating, fuel and water. You might also have supply of other media, such as steam or compressed air. Also try to get statistics on (or estimate) the production rate during the same period. Production data will enable you to estimate the specific energy use, for example in terms of kWh per produced unit, per hours worked or per tonne of raw material used. This can be useful both for allocating the energy and for calculating the result of suggested measures.

You can also ask the staff what activities are going on at the studied site, to get a picture of the unit processes and energy flow you should look for when you get there.

2.5.1. Electricity supply data

The electricity sales company and the electric grid owner are sources for the electricity supply data you need. It is good to have as much detailed data as you can when it comes to electricity use. The grid owner can provide consumption data with hour by hour resolution to their customers (at least in Sweden). If the object of the survey is small or if it has small sub-meters, there might be a problem to get hour by hour data. Energy data for a whole year, regardless of the time resolution, is recommended for analyzing the dependence on outdoor temperature.

Detailed data on the total electricity use during the period for which you are mapping the unit processes is also important. Sometimes you can use the customer ID number and a PIN code to get the hour by hour use from the supplier via the Internet. In that case it is easy to also download data for the period when you make your own electrical measurements. Another option is to log the total incoming electricity during the same period.

From the invoices you can get information on prices, power use, reactive power, etc. The invoice from January typically contains a summary of the energy use for the last year.

To some extent it is possible to analyze the data before visiting the company. You can compare the energy use graphs for production hours during different seasons to find electrical heating and cooling loads. You can also compare idling energy losses in different seasons and analyze total energy use, power peaks, reactive power, etc.

2.5.2. District heating supply data

Sometimes it is possible to get historical data on the energy used for district heating with hour by hour resolution, but often you have to be satisfied with the amount of heat used per month. Sometimes you can use the customer number and a PIN code to get data from the supplier via the Internet. If you ask for the service in advance, the supplier might be able to give you hour by hour resolution data for the time of the audit.

2.5.3. Fuel supply data

Fuels such as oil, wood pellets or wood chips are mostly bought in batches. To be able to calculate how much fuel was used in the past, you need to know how much fuel there was to start with, the purchased amount and how much is left. If the needed data is not available you might be able to estimate the fuel use by calculating the heat losses.

2.5.4. Other energy supplies

There are also other possible energy supplies, such as steam, compressed air, liquid nitrogen for cooling, welding gas, etc. As long as someone is charging for the supply you might be able to get some historical data, but it is not always the case. It is a good idea to look for informal energy flows: Two neighbouring companies can share the same district heating supply, air handling unit or air compressor, there might be solar heating for hot tap water or the company uses residues, such as wood chips, from the neighbour for heating. Energy supply not considered as energy can sometimes turn out to be important. Liquid nitrogen for cooling and welding gas for oxy-fuel welding are two examples.

2.5.5. Miscellaneous background data

Some other data that might be useful include:

- Water supply data, to be able to calculate the hot water use.
- Drawings of the building, floor plans.
- Drawing of the electricity supply systems.
- Drawing of other energy supply systems, for example hot water pipes, compressed air pipes
- Information on the ventilation: Drawings of ventilation ducts, number of air handling units and air flows.
- Information on time schedules for production processes, ventilation or lighting.
- Energy efficiency measures that are already taken.

Drawings can guide you to where in the system or in the building measurements can be made and what information to ask for. The drawings can also give you information that you can use directly. Air flows can be noted on ventilation drawings. Rated power for electric equipment, like motors and heaters, can be noted on detailed electrical drawings at the distribution board

http://dx.doi.org/10.5772/51717

level. When data such as documentation from a drawing is used, it is a good idea to make sure the data is up to date, by asking the staff about changes or by using measurements to validate.

2.6. Step 2: Visiting the company to collect data

By visiting the company during production hours, you can find out how the processes work, examine the design and function of systems and equipment, validate the descriptions, drawings and layouts you already have (for example of building and air handling units) and fill in the gaps where you have no data at all. When you visit the company you have the opportunity to:

- Speak to the key persons in different areas, operational and maintenance staff.
- Note the uptime and downtime for different devices, by registering how the processes work and by asking the staff.
- Inspect the building envelope and the space heating system.
- Review the procedures for service and maintenance.
- Go through all the unit processes systematically.
- Assess the function and condition of the equipment.
- Make an inventory of installed power, for example for motors, light sources and other devices.
- Measure the power used by a selection of important equipment. It might be the most energy demanding devices or any equipment you identify as critical to be able to divide the energy flow into different unit processes.
- Look for control systems (for the ventilation or for a production line). If there are any, you can try to find out if they are used and how they work. From the control systems for example you might get timetables with operation hours or logged data on production rate.
- Count the number of devices (for example light sources or computers) to make calculations later, using templates.

By speaking to key persons you can find out which unit processes are used at the company, uptime and downtime for different devices, operating hours for the factory and more. The key persons can be found among energy managers, production managers, caretakers, machine operators or others, depending on the organization. A caretaker or in-house electrician is a good guide when you want to find the devices you want to examine and the switchgear where to measure electricity. The production manager can inform you about the different production processes and their operating times.

Data on electricity use is mostly in focus. Data on airflows, water flows, pressure, temperature and size is often useful. Often you find set values and statistical data from sources like plans, technical documentation or monitoring systems. The main purpose of the survey is to get an overview - it is not necessary to measure all parameters during the survey to create the power and energy balance. Still, sometimes measuring is the easiest solution and more accurate and precise measurements might be necessary to be able to predict the results of proposed efficiency or conversion measures.

As the audit is a temporary project, the audit itself is not an incentive to install permanent meters. If sensors, for example, for power, current, air flow, pressure or temperature are already installed and relevant data already logged or is possible to log, of course they should be used for the audit.

Sometimes you can also log the output signal directly from the control system with a voltage or current logger, depending on the type of signal, mostly 0-10 V or 4-20 mA. The logged information can be used directly as indicator of the uptime of equipment, e.g., an air handling unit. For further calculations you also need to know what real condition the signal represents and adapt your logged values to that.

2.6.1. Electric power measurements

The data from the energy supply company is not sufficient for allocating energy use to all different unit processes. Internal energy measurements from the studied facility can sometimes be found, especially for well-monitored core production processes. Still you will probably have to measure the electric power used by a selection of important equipment to be able to allocate the energy use and construct the power and energy balance. To make electrical measurements is very often the main activity when the studied object is visited.

Apart from measuring the energy use directly with permanently installed meters, there are different ways to measure the power or current for electric equipment to help you calculate the energy use. To calculate the energy use from electrical measurements you will need the active power. When visiting the company, it is preferable to measure the active power directly with a wattmeter. Measuring the current and voltage separately will enable you to calculate the apparent power, which cannot be used directly for energy calculations. The apparent power is only useful when the power factor is known from previous measurements or from technical documentation. Rated power for different electric equipment can be used to some extent, but normally it should not be used for the calculations. Rather, it is more appropriate to obtain the power or current and power factor from measurements.

A short visit to the studied site will not provide sufficient information to allocate energy use. To see changes in power use, a combination of current transducers (AC current clamps) and digital recorders (data loggers) can be used. A useful method is to let the data loggers record the current for one week. This will give you information about changes in usage patterns during working hours as well as nights and weekends.

The use of electric energy is very often in focus in the energy audits conducted by Linköping University. Methods and material for electric measurements and evaluation of data are also developed for energy audit purposes. Examples are shown in section 3. Practical examples - Measurements and analysis.

http://dx.doi.org/10.5772/51717

2.6.2. *Air flow measurements*

Air flow measurements can be used to calculate the ventilation heat losses, and also to roughly estimate the electric power used for the fan motors. On the other hand, the electric power used for the fan motor can also give you an idea about the air flow, and the electricity is often easier to measure. Sometimes the optimal way to find out the operating time for a machine or the working hours in a building is to log the ventilation air flow or temperature or the electricity used by the fan motor.

2.6.3. *Temperature measurements*

Temperature data are particularly useful for calculating the energy use of the unit processes Space Heating and Cooling and Ventilation. Temperature measurements can be used to calculate the efficiency of heat exchangers for AHUs. To calculate the heat losses, the exhaust air temperature is crucial, but seldom registered by the monitoring system and even more seldom logged. To use standard values for efficiency of the heat exchanger is a useful shortcut.

Temperature logging can be used to identify operating time for equipment. A temperature probe can be attached directly to the surface of a machine or in an air flow or other medium connected to the studied equipment.

2.6.4. *Size measurements*

To analyze the heat losses from a building the size of the building and the properties of the building materials must be known. Measuring the size of windows and doorways, thickness of walls and height of ceilings is often part of the survey. For air flow measurements in ventilation ducts, the size of the duct must be known.

2.6.5. *Make an inventory of installed load and operating time*

A fast and effective means is to count motors, light sources, computers and other devices and note their rated power. You find the rated power for motors and other devices on name plates or in technical documentation. Also note the rated cos φ for the motors, for future loggings of electricity - if you don't get the opportunity to measure the true power factor or cos φ, at least you have an idea of the figure. Even though the scientific literature provides examples of this approach, see e.g. [18], it is strongly recommended that the power factor is obtained by measurements. This is due to the wide variation of power factor which occurs from different applications.

To get the energy use for computers, copiers, servers and printers, you can get a good enough estimate by counting the number of devices and calculating the power and energy used. You probably have to use templates and estimates for these.

To get the power and energy for light sources, a good method is to count the number of different light sources and note the different rated power, printed on the light source, on the package or in technical documentation. This method works as long as the lighting is controlled by an on/off switch only. When you have discharge lamps, such as light strips, you have to add the power used by the ballast (control gear). Make sure to find out if it is a magnetic ballast (very common, always in older installations but still used today) or the more modern electronic ballast. Magnetic and electronic ballast differ in power use. For light strips with the common magnetic ballast it means multiplying the rated power by 1,2 and for the more modern electronic ballast it means multiplying by 1,1.

The uptime for the lighting might be hard to find out. You can simply ask the staff, log the light or temperature at selected light sources or log current used for lighting to be able to see the true uptime and downtime.

2.7. Step 3: Visit the company during non-production hours

This step is performed to find out what equipment is operating during non-production hours and find out if it must be operating. You might find support processes - lights, air handling units, compressors - being switched on though there is nothing to support, but also production processes in idle or stand-by mode.

Eyes and ears are useful to find equipment running during non-production hours. The measurement methods used are the same as during production hours, but the focus may be on other processes.

2.8. Step 4: Create a power balance, energy balance and allocate energy use

When you start constructing the energy balance, you will have data with different quality – detailed production data, estimated operating time, rated data, etc. You will have data with different time spans and time resolutions - statistical data for a whole year, one week measurements, instantaneous measurements, etc. The data will also describe different parts of the energy flow - sometimes the energy supply, sometimes the usage and sometimes the energy loss. In Figure 5, an example of an energy balance is presented.

Below some different common cases related to allocating energy use are outlined.

Case 1 - All the supplied energy is allocated to one specific unit process. Sometimes you can find this phenomenon for support processes, such as fuels for internal transport or district heating for space heating. Some of the supplied energy can also be very specific for a certain production process, such as gas for welding or liquid nitrogen for cooling.

Case 2 - Almost all of the supplied energy is allocated to one specific unit process – all you have to do is to take something away. In Figure 5 this is the case for Space Heating and Hot Tap Water which are the only processes supplied by the oil boiler.

This can often be the case when you use a hot water system for both production and Space Heating and Cooling. One example is a district heating hot water supply used for both a washing machine for surface treatment and for the indoor climate and space heating system. The space heating is mostly weather dependent and varies with the seasons. The energy use of the washing machine totally depends on the rate of production. Given that you have at least monthly data from the district heating supplier and that you know the production rate during the summer, when the heating system is mostly turned off, you can estimate the energy use

http://dx.doi.org/10.5772/51717

for the washing. Then you can compare the energy use during the summer and winter, subtract the energy use for washing and assume the rest to be heat for space heating and air handling.

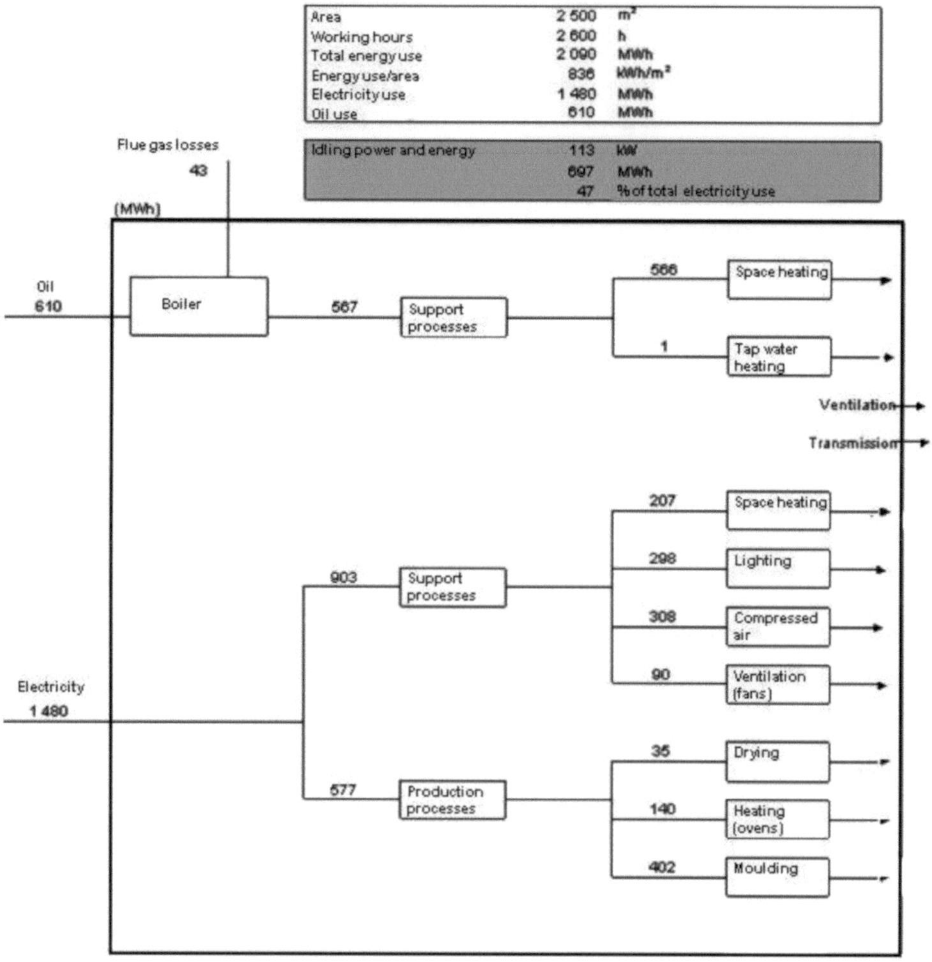

Figure 5. Energy balance for an audited plant [19].

Case 3 - Many different unit processes have the same supply, but at least some of them are running in a predictable way. This is often the case when it comes to electricity, as seen in Figure 7. From the switchgear or the distribution board, where you often have the practical possibility to measure, electricity is supplied to many different processes. Typically you find all the processes in one part of the building mixed. Even if the distribution board is totally dominated by one production process, like a laser cutter, you might also find support processes for that production process, like a compressor and a chiller, attached to the same board. The energy use for the laser cutter itself will vary a lot, but if the chiller works constantly as soon as the cutter is on and the uptime of the compressor already is measured for service reasons, you might be able to allocate the energy use by logging the total current to the distribution board.

If you have a common supply for even more processes, try to find the ones with constant power use, like light strips, or constant power use and programmed uptime, like air handling units with constant air volume and an autotimer.

Case 4 - A process has more than one energy supply. This is of course a common case. In a building with many air handling units, the units are all doing the same job – they are running the ventilation process. Even if there is a central remote control system for all air handling units, usually the energy use is not logged. The solution is to add up all the air handling units, accepting that your input data quality will range from rated power and estimated uptimes to detailed measurements.

You can also have one process with different energy carriers, e.g. a dryer consisting of an electric fan and a hot water heater. If you are not able to register everything, you might be able to use data from one of the carriers to calculate the uptime for the other. For example, the temperature of the heat exchanger can reveal if the fan is running or not.

You might also find two separate electricity supplies to the same unit process for safety reasons, i.e., two separate air compressors connected to two different switchgears to prevent stoppage. Of course you can have more than one supply to a single unit process for other reasons, for example if the electric distribution system has been supplemented because the process needs more electricity than the original switchgear was designed for.

2.9. Step 5: Creating an energy balance after adjustments (showing the impact of the suggested measures)

Depending on what kind of problems the auditor has found, different solutions are proposed. There are four principal means of reducing industrial energy costs as shown in Table 2.

Principal means of reducing industrial energy costs	Comment
Energy-efficient technologies	Improved efficiency among technologies using energy is one of the foremost and most common means of increasing energy efficiency in industry.
Load management	Reducing power costs by aiming to minimize the power loads is a common means for industry
Change energy carriers	Changing energy carriers, e.g. switching from oil to district heating is a means for industry to cut costs
Energy-efficient behavior	Energy-efficient behavior is a simple measure involving more efficient behavior among staff at the industry

Table 2. Four principal means of reducing industrial energy costs

http://dx.doi.org/10.5772/51717

It should be noted that the above table concerns energy costs. As regards reduced energy use, load management is not an energy-saving measure as it solely shifts the use of energy to times where it is more appropriate. Moreover, change of energy carrier may be a splendid way to reduce the use of non-renewable resources or energy costs. However, it may not lead to reduced energy use.

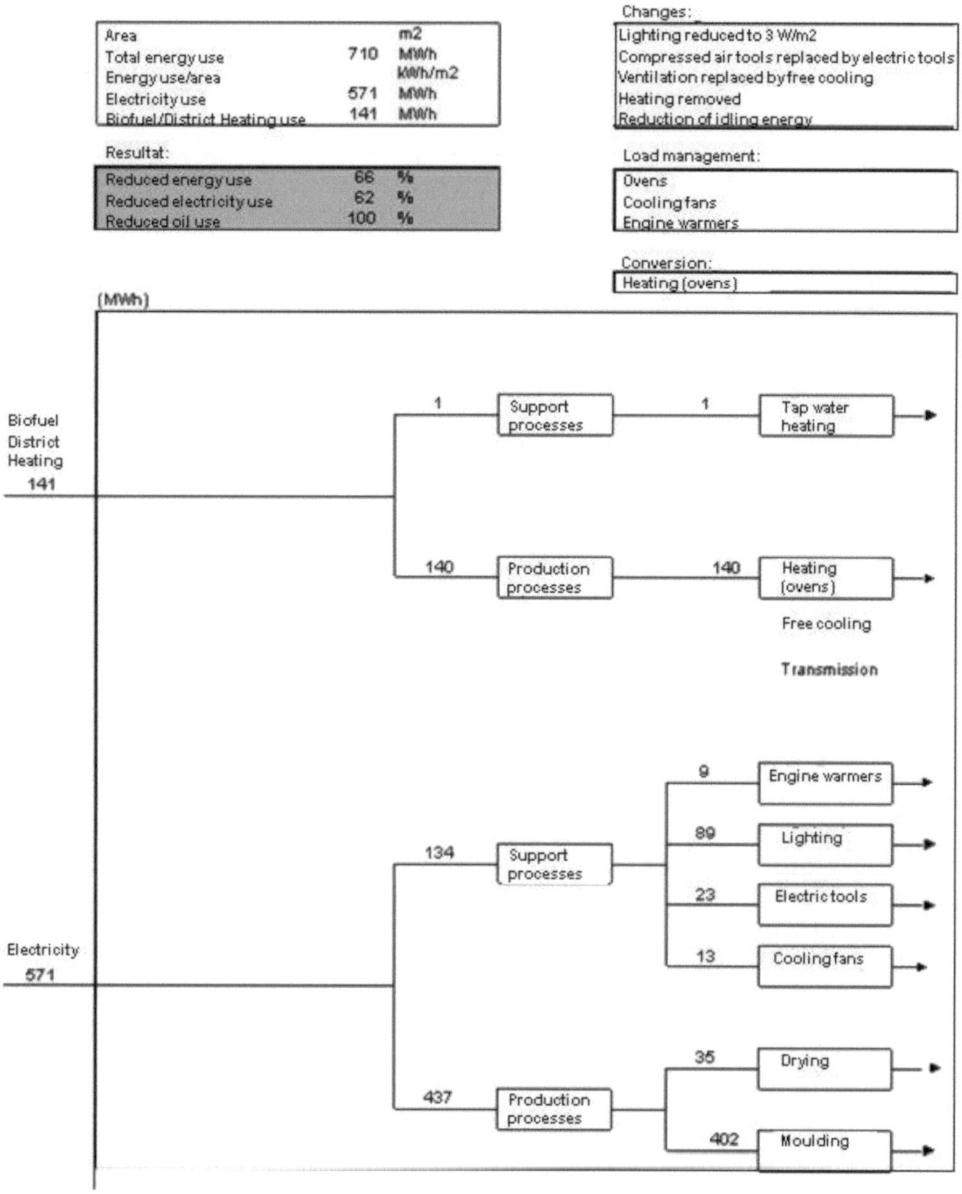

Figure 6. Energy balance after suggested adjustments [19].

When the impact of the proposed measures is calculated, a new balance is presented (Figure 6). The new balance can have two different purposes: It can show the potential for changes in the energy use or it can show the calculated impact of the measures. The potential for energy efficiency can be estimated with standard values for the best available technology. If the impact of the suggested measures is to be presented, more detailed technical and economical calculations are made.

2.10. Step 6: Confirmation of audit results

This part is of crucial importance, not least in regard to whether the energy audit results will, in fact, be accepted, and not discarded. Normally, a meeting is held with representatives from the industry concerned and the conductors of the audit about the proposed energy efficiency measures, and the analysis of the energy balance. In this step, the primary focus is on the proposed measures. For some processes, there are a number of different alternatives. This step enables the energy auditor and the industry representatives to discuss which measure is most suitable for the industry. Normally, this meeting is not with high level executives but is held with the maintenance manager or the energy controller.

3. Practical examples – Measurements and analysis

This section shows a few practical examples of measurements and analysis for energy audit purposes, focused on electricity. The building properties and heating and cooling systems are often important to examine in an energy audit. Air flow and temperature measurements are often needed. As there are fairly standardized methods for these tasks, they will not be further discussed in this chapter.

3.1. Allocation of electricity use – Measurement case description

In a typical measurement case we want to separate the electricity use for the unit processes air handling, lighting, compressed air, internal transports and packing. Instantaneous power measurements, current logging and rated power for equipment are used to allocate the electricity use.

Based on information from the staff at the industry, we know that the distribution board in Central Picking Stock is used for ventilation, lighting and electric tools for packing. The Central Garage distribution board is used for ventilation, lighting and a forklift charger. From the board Central Large Hall electricity for lighting, ventilation and packing equipment is distributed. Based on that information, and the drawings of the electricity distribution system in Figure 7, the plan for measurements may look like in Table 3.

The logging period is the same for all the current clamp meters, in this case seven days, to cover the variations in energy use in production hours as well as during the weekend. The range setting of the current clamp meters is to be appropriate for the load, but also depends on the type of clamps and loggers available at the moment. In this case we have 500 A, 200 A and 20 A current clamp meters.

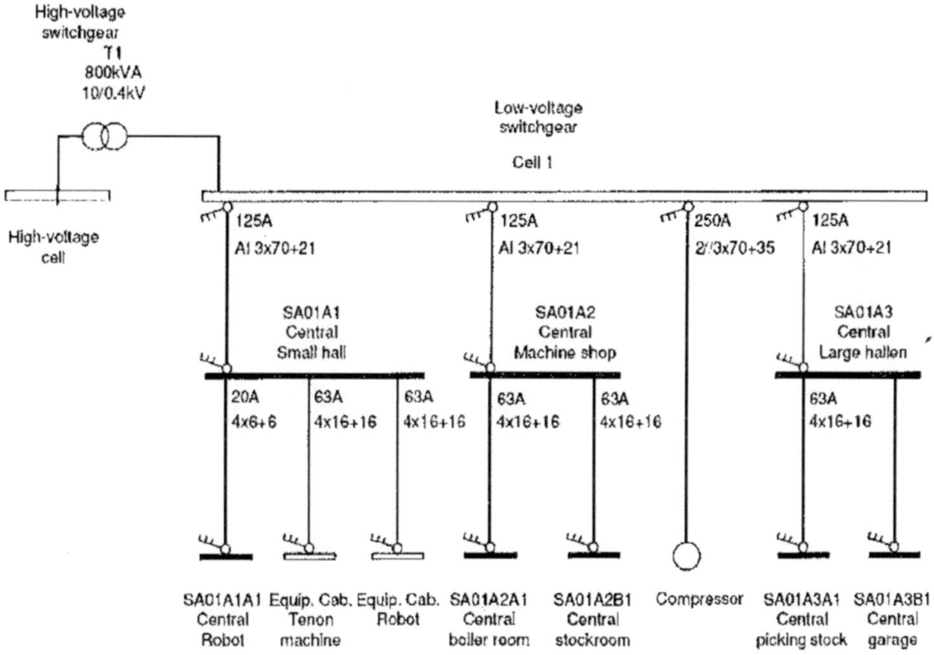

Figure 7. Example of a schematic drawing of an electricity distribution system.

The sampling interval for the current clamps can be changed depending on what we want to know. For the air compressor we may log more frequently, for example every five seconds, if we find that the compressor is switching mode very frequently, otherwise logging at 60-second intervals may be enough.

The purpose of the continuous measurement will also affect the logging interval. If the first visit gives us the impression that the compressed air system is an important part of the energy use, we may measure more carefully to have enough information to suggest detailed measures. The drawing of the electric distribution tells us that the fuse for the compressor is 250 A, but based on the information from the instantaneous measurements we plan to carry out, we might be able to replace the 500 A clamp with a 200 A clamp.

Unit process	Power data source	Time data source	Comments
Compressed air	Instantaneous measurements.	Logged data from current clamp 4.	Clamp 4 settings: range 200 or 500 A, sampling interval 5 to 60 s.
Lighting	Rated power, verified by instantaneous measurements.	Oral info on working hours. Verify with data from current clamp 1 and 2.	Clamp 1 settings: range 200 A, sampling interval 60 s. Clamp 2 settings: range 200 A, sampling interval 60 s.
Ventilation	Instantaneous measurements.	Air handling unit control system. Verify with data from current clamp 1 and 2.	
Internal transport	Rated power for forklift chargers combined with logged data from current clamp 3.	Logged data from current clamp 3.	Clamp 3 settings: range 20 A, sampling interval 60 s.
Packing	Rated power, instantaneous measurements.	Logged data from current clamp 1.	The residual energy is logged data from current clamp 1, minus clamp 2 and the energy for lighting and ventilation. The residual is equal to the energy for Packing.

Table 3. Example of plan for energy measurements to allocate electricity use to unit processes.

In this case we will have a measurement from a switchgear cabinet that covers several unit processes, for example lighting and forklift charging for a garage and picking stock, as shown in Figure 8. The data from clamp no. 1 in the switchgear must be complemented by simultaneous measurements from clamps 2 and 3 in the distribution board, where clamp 2 measures the total incoming current to the Central Garage board and clamp 3 measures the outgoing current to the forklift chargers.

Later, the data from clamp 2 in the Central Garage distribution board can be subtracted from clamp 1 in the switchgear to get a picture of the energy use in Central Picking Stock and Central Large Hall.

At the same time you have parallel measurements of the air compressor from clamp no 4. The clamp meter is placed in the switchgear, where the compressor has directly connected conductors, or in the compressor, depending on where you have the best access.

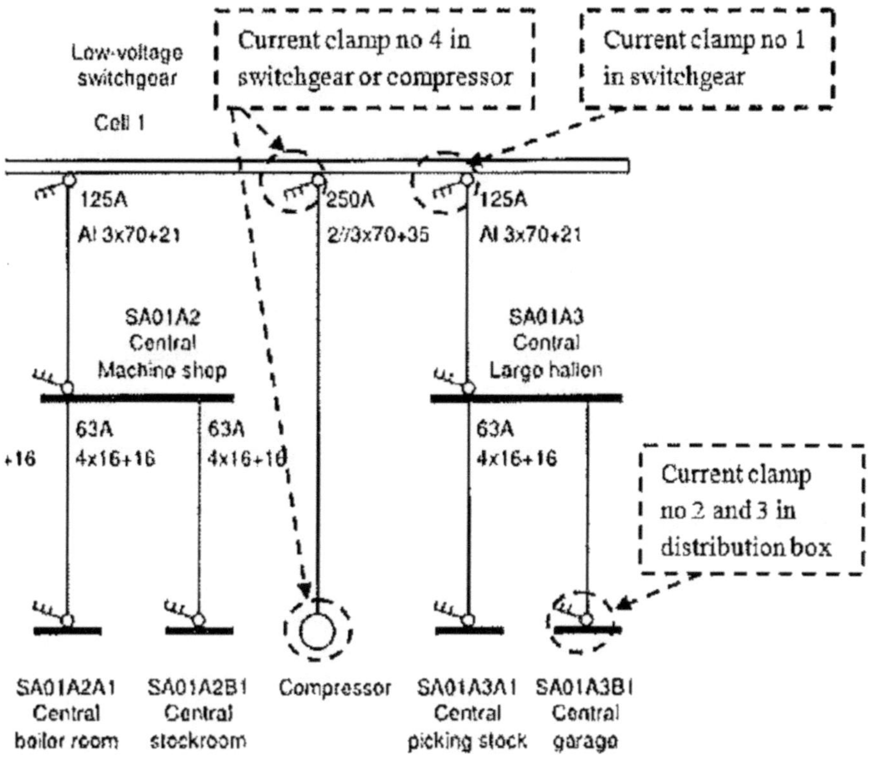

Figure 8. Part of a schematic drawing of an electricity distribution system, with locations for current clamps for simultaneous current measurement, to make it possible to subtract a part of the load from the total and separate electricity used for different unit processes.

3.2. Calculating the electricity use from logged data

Logged data can be used for calculating:

- Uptime.
- Average power for operating hours.
- Idling power.

In the example below the case is an hydraulic press, but the same procedure can be used for other equipment, for example to calculate the energy used to cover leakage in compressed air systems. With some more measurements and analysis, it can also be used for a complete building or workshop.

The idling energy, uptime and average power can be calculated in a few steps:

1. Measure the power factor at different loads.
2. Log the current.

3. Analyse the graph.

4. Convert the current to power.

5. Analyse the data again.

How to do this will be shown with measurements from a hydraulic machine for sheet metal processing.

3.2.1. Measure the power factor at different loads

The power factor is measured while the machine is running. The machine has two well-defined modes of operation – one when the metal sheets are actually shaped and one when the hydraulic motors are running without any load, i.e., idling.

Mode	Power (kW)	Power factor, λ
Active	< 250	0,82
Idling	48	0,38

Table 4. Measured power and power factor for a hydraulic press.

3.2.2. Log the current

The current is logged for about a week, to cover working days as well as the weekend. The original current logging data is presented in a graph (Figure 9).

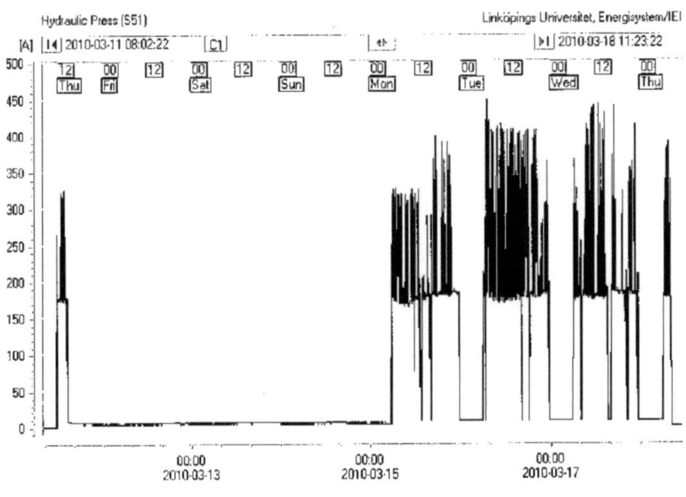

Figure 9. The current used by a hydraulic press during one week.

3.2.3. *Analyse the graph*

From the graph (Figure 9), it is clear that the machine is not used Friday to Sunday and for some hours during the nights. The machine is turned off during non-production hours. Nonetheless it has some idling power during the weekend and nights. The conclusion is that there are in fact three different modes to consider: active, idling and standby. The periods with zero current in the beginning and the end of the graph is the time before the logger was connected to the machine and after it was disconnected.

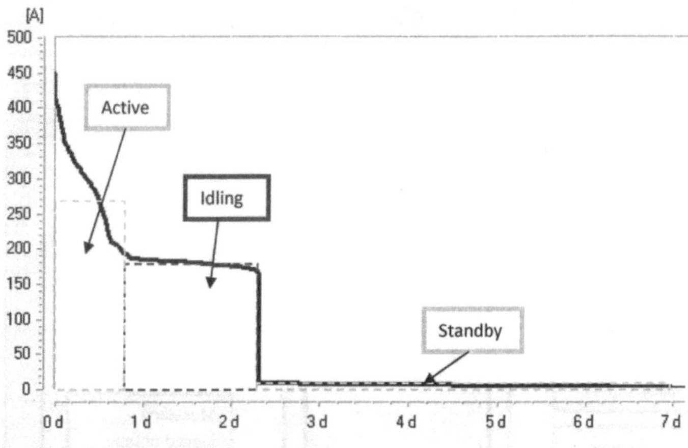

Figure 10. Duration diagram of the current used by a hydraulic press during almost one week (6,8 days). The active mode lasts for about 20 hours, the idling mode for about 1,5 day and the standby mode for about 4,5 days.

One way to analyse the graph is to create a duration diagram with the logged data, as in Figure 10. The duration diagram can be used to distinguish between the modes and to estimate the energy use for the different modes.

3.2.4. *Convert the current to power*

The power can roughly be estimated with the duration diagram together with the instantaneous measurements of the power factor.

Mode	Power factor, λ	Current (A)	Voltage (V)	Power (kW)	Duration (h)	Energy (kWh/week)
Active	0,82	270	230	150	20	3000
Idling	0,38	180	230	47	36	1700
Standby	?	5	230	3,5	108	400
SUM						5100

Table 5. Power and energy use for a hydraulic press during one week.

In this case a close calculation based on the raw data gave a similar result (5176 kWh) to the rough estimate, but that is mostly because the same assumptions were used. Changing the duration or average power just a little affects the result a lot. The power factor at partial load and at standby load is also unknown, which affects the accuracy of the calculations.

3.2.5. Analyse the data again

Together with information about the rate of production and the working hours, the annual energy use for this equipment can be calculated.

If this is a representative production week, the result can simply be multiplied by the number of working weeks per year. In this case it is not a typical week but information from the production manager tells us the working days are representative. The average power use during working hours can be calculated and multiplied by the number of working hours per year. In this case the rest of the year can be considered as standby time. In other cases equipment can be completely turned off during longer holidays.

Mode	Power, average (kW)	Duration (h/year)	Energy (MWh/year)
Working hours (active and idling mode)	84	3456	290
Standby	3,5	5304	19
SUM		8760	309

Table 6. Calculated annual energy use for a hydraulic press.

We know from previous calculations that about 35 percent of the energy use during working hours is really idling energy. That information is not needed for the allocation of the energy use to the right unit process, but it is important information for further analysis and to suggest measures. A closer look at the graph of the current used during a week also reveals that the machine sometimes is idling for more than one hour without producing (see Figure 9). One example is Monday before midnight, when it is idling for about two hours before it is at last turned off for the day. An automatic control system or changed routines for the operator can be used to switch to standby mode faster. The aim of the analysis was to be able to calculate the energy use, but on the way some possible measures were also found.

4. Concluding discussion

This chapter has presented the energy audit methodology being developed at the Division of Energy Systems at Linköping University for a period of more than 30 years. Despite extensive research and experience in the field – about 500 energy audits have been carried out based on this methodology - the energy audit methodology has not been explicitly outlined previously. This chapter may therefore still contain areas for improvements. This chapter has for example

http://dx.doi.org/10.5772/51717

not covered in depth the obvious risk of errors when collecting and analyzing the data and results in an energy audit. For practical reasons, measurements are normally carried out for one or two weeks. When setting up an energy balance, this period of time is normally assumed to be a representation of a "normal" week, and thus multiplied by a factor in order to gain an annual balance.

Moreover, it must also be mentioned that this methodology covers a Swedish and Scandinavian context, for which reason some material might be difficult to adapt in countries outside of Scandinavia. Finally, this chapter has not explicitly covered the in-depth calculations needed for various parts of an energy audit as there is no room for that in this context. The chapter does not cover all the various aspects of the energy audit methodology but rather gives the reader a basic introduction to the subject. After all, becoming a good energy auditor is not about reading a book, but is rather a continuous learning process where experience, not theory, is the major part of increasing skills.

Author details

Jakob Rosenqvist*, Patrik Thollander, Patrik Rohdin and Mats Söderström

*Address all correspondence to: jakob.rosenqvist@liu.se

Department of Management and Engineering, Linköping University, Linköping, Sweden

References

[1] European Commission(2006). Directive 2006/32/EC of the European Parliament and of the Council of 5 April 2006 on energy end-use efficiency and energy services and repealing Council Directive 93/76/EEC, Brussels.

[2] Schleich, J. (2004). Do energy audits help reduce barriers to energy efficiency? An empirical analysis for Germany. International Journal of Energy Technology and Policy , 2, 226-239.

[3] Caffal, C. (1996). Energy management in industry. Centre for the Analysis and Dissemination of Demonstrated Energy Technologies (CADDET). Analysis Series 17. Sittard, The Netherlands.

[4] IPCC(2007). Contribution of Working Group III to the Fourth Assessment Report of the Intergovernmental Panel on Climate Change. Summary for Policymakers. http://www.ipcc.ch/SPM0405pdf (accessed October 8, 2007).

[5] IEA (International Energy Agency)(2007). Key world energy statistics 2007, Paris. from: http://195.200.115.136/textbase/nppdf/freekey_stats_2007.pdf (accessed January 8, 2008).

[6] ECON(2003). ECON centre for economic analysis AB. Konsekvenser på elpriset av införandet av handel med utsläppsrätter [Impact on price of electricity of the introduction of trading with emission rights]. Ministry of Industry, Employment and Communications, Stockholm. [in Swedish]

[7] Clara Pardo Martinez (2010). Factors Influencing Energy Efficiency in the German and Colombian Manufacturing Industries, Energy Efficiency, Jenny Palm (Ed.), 978-9-53307-137-4InTech, Available from: http://www.intechopen.com/books/energy-efficiency/factors-influencing-energy-efficiency-in-the-german-and-colombian-manufacturing-industries

[8] Patrik Thollander, Maria Danestig and Patrik Rohdin, Energy policies for increased industrial energy efficiency- Evaluation of a local energy programme for manufacturing SMEs, (2007). Energy Policy, (35), 11, 5774-5783.

[9] Johansson, B, Modig, G, & Nilsson, L. J. (2007). Policy instruments and industrial responses- experiences from Sweden. In: Proceedings of the 2007 ECEEE summer study "Saving energy- just do it", Panel , 7, 1413-1421.

[10] SEA (Swedish Energy Agency)(2006a). Energy in Sweden 2006. Swedish Energy Agency Publication Department, Eskilstuna.

[11] EEPO(2003). Year 2002 Results. European Electricity Prices Observatory (EEPO), Brussels.

[12] Lytras, K, & Caspar, C. (2005). Energy Audit Models. SAVE-project AUDIT II, Topic Report.

[13] ASHRAE (2003). Application handbookAmerican Society of Heating, Refrigerating and Air-Conditioning Engineers, Atlanta.

[14] Mazzucchi, R. P. (1992). A guide for analyzing and reporting building characteristics and energy use in commercial buildings, ASHRAE transactions , 98(1), 1067-1080.

[15] Nilsson, P. E. ed. (2003). Achieving the desired indoor climate- Energy efficiency aspects of system design. Commtech group, Lund: Studentlitteratur.

[16] Thollander, P, Mardan, M, & Karlsson, M. (2009). Optimisation as investment decision support in a Swedish medium-sized iron foundry- a move beyond traditional energy auditing. Applied Energy, 86(4), 433-440.

[17] Franzén, T. (2005). EnergiSystemAnalysMetod, EnSAM, för industriella energisystem [Energy System Analysis Method, EnSAM, for industrial energy systems]. 2005, Linköping Institute of Technology, Lin-köping. [in Swedish]

[18] Saidur, R, Rahim, N. A, Ping, H. W, Jahirul, M. I, Mekhilef, S, & Masjuki, H. H. (2009). Energy and emission analysis for industrial motors in Malaysia. Energy Policy, 37(9), 3650-3658.

[19] Trygg, L, & Karlsson, B. (2005). Industrial DSM in a deregulated European electricity market- a case study of 11 plants in Sweden. Energy Policy, 33(11), 1445-1459.

Chapter 6

The Role of Building Users in Achieving Sustainable Energy Futures

Tim Sharpe

Additional information is available at the end of the chapter

http://dx.doi.org/10.5772/51900

1. Introduction

In the drive to provide a sustainable energy strategy the reduction of energy use by buildings is a crucial component as they provide the majority of energy use and carbon emissions. In an attempt to mitigate the damaging effects of greenhouse gas emissions, international governance has legislated for the reduction of energy use and CO2 emissions. In Scotland (the setting for this research) the Government has identified target reductions in domestic regulated energy use of 30% by 2010 and 60% by 2013 (compared to 2007 technical standards) and the ambition of whole life zero carbon by 2030 [1]. A low carbon economy is now a strategic priority for the Scottish Government. As domestic energy use represents 30% of total national energy use [2] there can be little doubt over the role this sector has to play in helping to achieve the targeted reductions. Whilst for new buildings this may be addressed through building standards, a more pressing problem is that an estimated 70% of the stock currently in existence will still be standing and in use by 2050, and much of this stock has a very poor performance [3]. Therefore the role that existing dwellings will have to play in helping to meet these ambitious targets cannot be underestimated.

The primary mechanism to affect change has been improved building regulations, for which compliance is achieved at design stages. Although standards have improved significantly in recent years, it is becoming increasingly apparent that these are not being translated into energy savings in practice [4]). In situations where thermal improvements are made there is emerging evidence that the drive for energy reduction is resulting in other unintended negative consequences, for example poor indoor air quality which as well as being problematic in its own right, also leads to rebound behaviours which undermine energy strategies [5].

Problems of energy consumption and carbon emissions apply to both new and existing buildings. Existing buildings are in many ways a more significant problem, in that they tend to have very much worse energy performance; make up a much larger proportion of the stock; can have physical, economic and cultural barriers to major improvements; and are not subject to the same regulatory requirements as new building as current building standards are not applied retrospectively.

The use of Building Performance Evaluation (BPE) is a crucial tool in assessing the tangible performance of buildings, and identifying the positive and negative factors that lead to actual consumption. The Mackintosh Environmental Architecture Research Unit (MEARU) has been at the forefront of developing and promoting forms of BPE [6] and has undertaken a range of evaluations in both new build and existing buildings. BPE includes both qualitative and quantitative methods to gather data on energy use, environmental performance and occupant behaviour and attitudes. From this is it possible to identify actual energy consumption, patterns of occupancy and behaviour, and the environmental conditions that are being achieved and from these determine process changes in design, management, procurement, construction and use that can improve building performance. The use of BPE is crucial to sustainable urban futures is as it identifies the gaps that occur between design, construction and occupancy. Wider use of BPE in the future may place more of an onus on designers to consider actual performance, as opposed to designing for regulatory compliance. It is a technique that can be applied to both new build and existing buildings.

Of these, the issues of occupancy are attracting the most interest. The potential impacts of occupant behaviour on energy consumption are significant, with some studies identifying variation in consumption by a factor of 4 and 5 times between identical dwellings [7]. Of equal importance however is the question of the impacts *on* occupants of low energy design in respect of environmental performance, especially indoor air quality, and what the implications are for energy consumption and health.

This chapter will describe and compare two case study projects that have used BPE to investigate performance in use, as a comparison of two very different building types. The first of these is the refurbishment of a 19th century Grade A listed tenement building in Edinburgh; the second is the 'Glasgow House' a prototype low energy housing development for Glasgow Housing Association.

The former is an existing 19th century stone built tenement in Edinburgh's Grassmarket that was refurbished to a high standard, including improved fabric performance through internal insulation and secondary double glazing, sun spaces, a ground source heat pump supplying underfloor heating, and a mechanical heat recovery ventilation system (MVHR) system.

The 'Glasgow House' is a new build project developed by Glasgow Housing Association (GHA), one of Europe's biggest landlords, as a prototype for future housing developments in the city. The design proposed a thermally heavy clay block system, with high thermal performance, glazed sun spaces, MVHR, solar thermal hot water heating; high efficiency gas boiler and low energy lighting equipment. Due to uncertainties about this type of construction, two test houses were constructed by GHA's partner organisation, City Building one of which uses a more standard form highly insulated timber frame.

Although very different house types, built over 120 years apart, both are attempting to meet contemporary standards in terms of energy use. In the evaluation several common factors relating to ventilation and indoor air quality (IAQ) were apparent in the performance of both projects and the question that this chapter addresses is how these factors affect building occupants, their subsequent behaviour, and how this in turn affects energy consumption.

2. Indoor air quality

IAQ is an important, but neglected aspect of sustainable design, which more commonly emphasises energy use and carbon reduction. However, achieving good IAQ is important for a number of reasons. Firstly it is crucial for health and well-being of occupants. Secondly, it is increasingly evident that poor IAQ can lead to detrimental energy performance, for example, users opening window to control temperature, humidity, stuffiness and smells, even when mechanical systems are intended to address these issues. Thus the tension that exists between low energy design, which attempts to minimise ventilation loss, and good IAQ, which seeks to maximise ventilation, needs to be addressed.

The majority of the world's population spends 90% of their lives indoors [8], [9]. Its quality is of recognized concern [10] and can be affected by many factors, most noticeably air temperature (Ta), as well as surface temperature (Ts), humidity and pollution levels. IAQ affects how inhabitants perceive a space, to the same extent as the availability of space and light do. Through sound and well tested ventilation design a healthy living environment can be achieved.

Globally, indoor pollution has been related to respiratory illnesses [11]; has resulted in an increase in childhood asthma [12] and poor levels of IAQ have been linked with mechanical ventliation and sick building syndrome [13]. Factors that contribute to IAQ can be considered in various ways and calculated using different indicators. Allard defines optimum IAQ as,

"...air which is free from pollutants that cause irritation, discomfort or ill health in the occupants." [14].

Scottish Building Standards (SBS) states that indoor air quality should not endanger the health of the inhabitants [15]. It suggests a temperature range of 18 - 21°C, relative humidity (RH) of below 70% as well as specifying trickle vent sizes to maintain air quality. Although clearer than the previous definition, the standards expediency is debatable, producing only the minimum levels of IAQ needed, whilst focusing on maximising energy efficiency [16]. Temperature and RH ranges are not room specific, and with indoor pollution varying over time, the advised levels of ventilation should be adaptable [17].

CO_2 is an appropriate indicator to measure when assessing IAQ and was used in these studies as its importance as an environmental indicator is invaluable. The concentration of CO_2 is very rarely found at hazardous levels indoors, but levels of CO_2 represent the presence of other contaminants in the air, such as bio-effluents, which relate directly to health issues [18]. Increased levels of CO_2 are indicative of occupancy and inadequate ventilation [19]. Pettenkofer first tested air for the presence of CO_2 [20]; consequently Pettenkofer's Max, of 1000ppm, was established and the current consensus of opinion is that levels above 1000ppm are linked to poor occupant health [21]. Where concentrations greater than 1000ppm are experienced the rate of air change is insufficient and the potential for culmination of internal pollutants is increased with an associated impact on occupant health. Examples within domestic contexts include volatile organic compounds (VOC), which act as allergens and respiratory and dermal irritants [22].With low air change rates there is also a well-defined risk of interior moisture vapour build up which brings with it its own set of health implications. Vapour pressures over 1.13kPa have been identified as promoting the growth of dust mite populations [23] which have, in turn, have been found to have a causal relationship with development of asthma in susceptible children [24]. With high vapour pressures there is also an associated risk of fugal growth and an increase in the levels of fungal spores, microbial bodies and other pathogens which can be detrimental to the health, particularly to the ever increasing atopic portion of the population. In addition to this, increased relative humidity has also been found to increase health impact from non-biological aerosols as it increases the rate of off gassing of water-soluble chemicals such as formaldehyde [25].

In both these case studies a key component to address the issue of ventilation and energy use was the inclusion of Mechanical Ventilation Heat Recovery Systems (MVHR). The principle of these systems is that poor quality, but relatively high temperature internal air is mechanically extract from spaces in the dwelling, typically spaces which contain 'problem' air such as high moisture content or smells from kitchen and bathrooms. This air is passed through a heat exchanger during which colder fresh air from the outside is warmed by the recovered heat before being delivered to the dwelling. In theory should satisfy the needs of both energy conservation and IAQ to produce a sustainable solution. Accordingly the discussion below makes particular reference to the performance of these systems.

3. BPE methodology

The methodology in both case studies was broadly similar. Quantitative data on temperature, humidity and CO_2 levels was collected using Eltek GD-47 Transmitters linked to Eltek RX250AL1000 Series Squirrel Data Loggers. This was supplemented by Gemini TinytagPlus Data Loggers for temperature and humidity some rooms without a power supply (bathrooms and toilets).

The Glasgow House was unusual in that as demonstration houses they did not have occupants. MEARU in conjunction with GHA developed a methodology for scenario testing whereby volunteers occupied the houses for two-week periods during which they were asked to follow set occupancy 'scripts' for behaviour. Heating and environmental controls were fixed in the script and users were asked not to change these. Thus occupancy and behaviour could be tightly controlled, allowing an examination of the environmental performance under known conditions

In the Glasgow House, additional qualitative information was gathered using occupant diaries, record sheets for key activities such as fan operation and boost switch use, cooking, and window opening. The inhabitants of the house were each given diaries to record their day-to-day activities. This included the documentation of house occupancy periods, personal sanitary routines, and the use of individual electrical equipment. Cooking periods and kettle use were recorded separately, as well as instances when the boost switch was used on the extract for the MVHR system, usually in association with showering or food preparation. Room occupancy levels and window opening was also documented. Participation in post occupancy evaluation (POE) questionnaires allowed for the occupants qualitative and functional responses towards the houses to be gathered.

In Gilmores Close, due to the more vulnerable nature of the occupants (a high proportion of which have special needs), a semi-structured interview was conducted with residents and office users to query patterns of occupancy, user behaviour and comfort. This was supplemented by surveyors observations, photographs and thermographic imaging

4. Case study 1: The Glasgow house

4.1. Construction

The two semi detached houses were built in 2010 by City Building (CB) in partnership with the GHA. They were designed to provide comfortable and flexible living for low income families with an aim of costing no more than £100 per year to heat. The houses are of similar layout and consist of a porch, kitchen/dining/living area, a utility room/WC, four bedrooms, a bathroom and garden.

The design incorporated high levels of thermal efficiency using a Thermoplan clay block with external insulation, highly insulated roof cassettes and high performance windows, thermal mass, airtight construction (0.4 ach), sunspaces, solar thermal hot water collectors, mechanical ventilation heat recovery, low energy lighting and high efficiency appliances (House A). The comparison house is identical house except the clay block is replaced with a more conventional highly insulated timber frame system which is the standard form of construction used by City Building.

A series of scenarios were proposed to examine how these houses performed in use with actual occupants. Findings from two separate periods of BPE, in February and December 2011 are described here. In both cases four occupants inhabited each dwelling. They were to simulate an average family living pattern and were given prescribed scripts on occupancy.

Heating is by a high efficiency gas boiler and radiators. Hot water heating is supplemented by the use of a solar hot water system. Ventilation is by a MVHR system, extracting air from the kitchen and bathrooms spaces and supplying a balanced flow of air to the living room and bedrooms. The temperature of the house was pre-set using the main thermostat and by the individual thermostats on the radiators in each room. The MVHR system was inspected visually and the filters changed if necessary. Electricity and gas meter readings were made at the start and end of the monitoring process in each dwelling to record energy consumption.

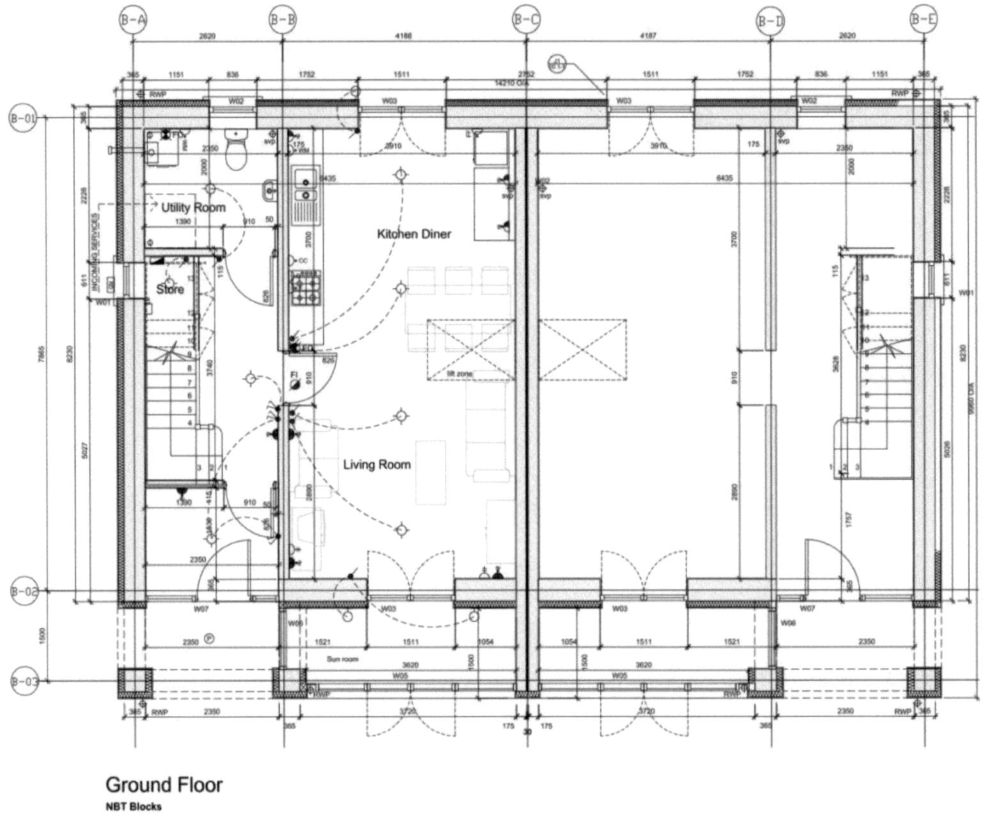

Figure 1. The Glasgow House, Ground Floor Plan

4.2. Data review

When assessing IAQ in relation to the specified Ta, RH and CO_2 criteria, the maximum and minimum values recorded are of most significance as well as the subsequent range produced. A graphical output, visually representing the continuous change in these values over time is appropriate for analysis and discussion. Mean values calculated from these indicators can sometimes be of use to give an overall representation of the data. However, this statistic can overlook significant individual moments, diluting the importance of some data, and thus failing to give an accurate depiction of performance. Mean values may be more appropriate for other IAQ indicators, such as VOCs and plasticisers, which have constant background levels. The data collected was subject to certain variables and limitations.

4.2.1. Occupancy

The first study period in February (SP1) ran from 15/02/11 at 00:00 until 27/02/11 at 23:55. In December study period two (SP2), ran from 06/12/11 at 00:00 to 15/12/11 at 23:55. Ideally both study periods would be for the same duration, however, it was determined that the se-

http://dx.doi.org/10.5772/51900

lected periods (SP1 and SP2) from each study were sufficient to provide adequate data for analysis and comparison.

The February study monitored a period of non-occupancy prior to SP1 commencing. This data provides a valuable 'control' period (CP) which can be utilised to understand how the house performed in relation to IAQ when uninhabited. Although the studies were run during different months, they were both in the winter period of the same year. Average external temperatures during this period were 5.5 ^{0}C in SP1 and 3.3 ^{0}C in SP2. The studies can be compared reasonably accurately in relation to this slight inconsistency.

During study periods occupancy levels in the houses remained constant. The houses' flexible layout resulted in sleeping arrangements varying depending on how the show home had been set up. (Table 1).

	House A		House B	
	Type	**Occupancy**	**Type**	**Occupancy**
Bedroom 1	Double	1	Double	1
Bedroom 2	Double	1	Double	1
Bedroom 3	-	0	Single	1
Bedroom 4	Twin	2	Double	1

Table 1. Occupant Sleeping Arrangements, SP 1 & 2, House A

When assessing the results it should be noted that in House A, Bedroom 3 was set up as a study/office and was not used for sleeping. As a consequence, the occupancy of House A, Bedroom 4 was double that of House B. Data collected from House A, Bedroom 3 is still relevant, although the results will not warrant accurate comparison with House B, Bedroom 3. The data can be viewed to see how the room performs when uninhabited, similar to the control period mentioned previously. House A, Bedroom 3 can be compared between the two study periods, however. It is worth mentioning that each occupant will not have spent the same amount of time in their bedroom and sleeping patterns will have varied. Similarly, room occupancy level throughout the house may have varied from time to time, depending on occupant activities and interaction. High occupancy room levels ≥ 3 inhabitants.

When evaluating the data it is essential to consider each bedrooms qualities, in order to make fair comparisons. (Table 2).

The room on the 2nd floor, in the attic space, is the largest of the bedrooms. Bedroom 3 is the smallest. Bedroom sizes vary slightly between the two houses. This is due to the construction types, which alter the wall build up, affecting internal space a little. The houses have identical plan configurations.

		House A		House B	
	Location	**Floor Area (m²)**	**Volume (m³)**	**Floor Area (m²)**	**Volume (m³)**
Bedroom 1	1st Floor	11.89	28.41	12.02	28.74
Bedroom 2	1st Floor	11.89	28.41	12.02	28.74
Bedroom 3	1st Floor	7.33	17.52	7.28	17.41
Bedroom 4	2nd Floor	18.69	39.43	18.69	39.43

Table 2. Bedroom Information

The MVHR system was serviced and its installation altered between SP1 and SP2.The studies had prescribed occupancy patterns to achieve specific goals and were scripted to accurately represent the airtight dwellings' IAQ performance using the MVHR system. The occupants were asked to refrain from naturally ventilating the dwellings, by opening windows and doors, and not to alter temperature thermostats. However, during SP1 windows were opened by the occupants in both House A and House B. (Table 3).

SP1 - House A	No. of Openings	Total Duration of Openings (min)
Bedroom 1	4	225 (3hr 45min)
Bedroom 2	7	1779 (29hr 39min)
Bedroom 3	0	0
Bedroom 4	2	578 (9hr 38min)
Total	13	2582 (43hr 2min)
SP1 - House B	**No. of Openings**	**Total Duration of Openings (min)**
Bedroom 1	1	33
Bedroom 2	2	215 (3hr 35min)
Bedroom 3	0	0
Bedroom 4	0	0
Total	3	248 (4hr 8min)

Table 3. Frequency of Window Opening, SP1, House A and B

The consequence of this natural ventilation will have more of an impact on the data collected in House A, simply due to the greater over all duration in which the windows were open, than in House B; 43 hours compared with 4 hours, respectively. Conclusions drawn from the particular data in both houses should acknowledge these variances in terms of their effect on Ta, RH and CO2 levels within the specified rooms.

During SP2 window opening was more tightly controlled and no natural ventilation was recorded, however, an occupant in House B increased the radiator thermostats in the kitchen, living room, utility and attic bedroom from one to four, over a period of 18.75 hours on 06/12/11. This change may have affected the data collected for House B, in particular in Bedroom 4. Any conclusions drawn from the particular data should take into account this variance.

Whilst this project is focused primarily on investigating CO_2 levels, it is worth including the analysis of the other monitored indicators, Ta and RH. Observing several IAQ components provides a greater understanding of how the MVHR system is functioning. Over the study periods all three of the indicators contributed to unhealthy IAQ within the bedrooms. CO_2 levels were of significant concern.

4.2.2. Air temperature data

The graphical output produced, as well as the statistics calculated, shows that Bedroom Ta fell out with the preferred parameters, defined as 18-20^{0}C, during both studies. The houses failed to sustain a constant bedroom Ta, within the optimum 2^{0}C range. The temperature related IAQ was unsuitable for sleeping on occasions. (Table 4) Within the data, an apparent difference is visible between the SP1 and SP2 results. Although ranges are similar, SP2 bedroom Ta was lower than the previous study, resulting in a reduced overall Ta. House A and B appear to function very similarly to one another in the studies.

Bedrooms	Max (°C)	Min (°C)	Range	Mean (°C)
SP1 – House A	23.30	16.60	6.70	20.01
SP1 – House B	22.40	16.70	5.70	19.82
SP2 – House A	20.40	14.50	5.90	16.95
SP2 – House B	19.30	13.50	5.80	16.25

Table 4. Overall Bedroom Ta Statistics

4.2.3. Relative humidity

Similarly, RH failed to maintain the standard defined as 40-70%. With corresponding to Ta values, maximum RH values are within the prescribed range, but the minimum values recorded fall below. This resulted in the occupants experiencing reduced IAQ during these intervals. (Table 5)

Bedrooms	Max (%)	Min (%)	Range	Mean (%)
SP1 – House A	59.30	31.10	28.20	43.37
SP1 – House B	67.50	32.50	35.00	43.25
SP2 – House A	56.40	35.30	21.10	43.96
SP2 – House B	61.20	36.90	24.30	45.40

Table 5. Overall Bedroom RH Statistics

There appears to be little difference in RH between SP1 and SP2. Also, House A's performance does not differ from that of House B.

4.2.4. Carbon dioxide data

Both study periods reveal comparable trends in CO_2 levels, highlighting a strong diurnal range. It is also clear that there is an identifiable difference between the general CO_2levels in SP1 compared with SP2. Comparing identical bedrooms in House A and House B is also of interest.

Bedrooms	Max (ppm)	Min (ppm)	Range	Mean (ppm)
SP1 – House A	2007.00	478.00	1529.00	872.50
SP1 – House B	2006.00	445.00	1561.00	921.06
SP2 – House A	1300.00	367.00	933.00	605.26
SP2 – House B	1478.00	375.00	1103.00	633.82

Table 6. Overall Bedroom CO2 Statistics

In SP1 CO_2concentration reaches levels double the maximum recommended value highlighting unhealthy IAQ. SP2 produced healthier results; however maximum values in each house still rose above Pettenkofer's Max. (Table 6) In general, bedrooms in House A and House B appear to function similarly in relation to CO_2.

Figure 2. Comparison CO_2Levels, SP1, House A, Bedroom 2 & Living Room

The diurnal range is clearly visible on the graphical information produced for the study periods. (Figure 2) This day to night change can be attributed to the bedrooms occupancy pattern, prescribed by their function. A graph showing a 24 hour period within a bedroom allows the rise in CO_2levels to be identified as occurring during the night, and can be associated with periods of sleep, and therefore occupancy. When CO_2 levels are lower the room is most likely to be empty because respiration is not taking place. Occupancy patters within the living room, also identifiable through CO_2 levels, are quite different to that of the bedroom.

http://dx.doi.org/10.5772/51900

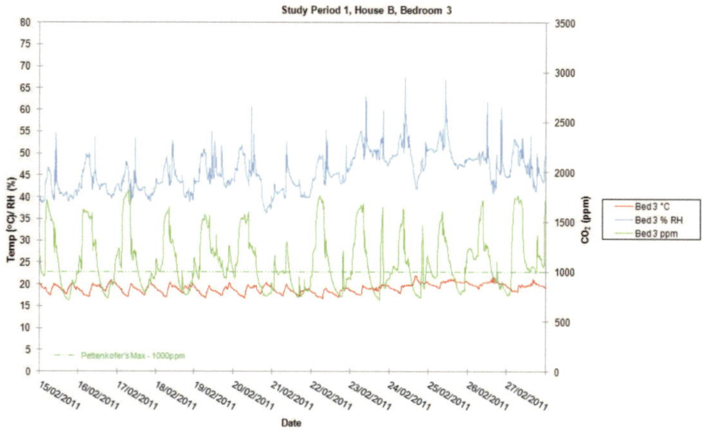

Figure 3. Unhealthy Bedroom Performance, SP1, House B, Bedroom 3

The most unacceptable levels of CO_2 found by the studies were recorded during SP1 in House B, Bedroom 3. (Figure 3) There was one occupant sleeping in this room. Levels of CO_2reached over 1000ppm for approximately 13 hours of each day, more than half of the total study period. The maximum level recorded was 1819 and the minimum being 713ppm (Table 7). The range of 1106 is evidence of the occupancy pattern, however, the diurnal range is less defined in this instance and may show that the room was also occupied at times during the day. When the room was in use CO_2levels in the air indicated a potentially harmful IAQ. When discussing IAQ, Ta and RH data recorded for this example support poor conditions, with max. and min. of 21.9°C and 16.7°C, 67.5% and 36.4% respectively. There is strong instability evident within all three indicators.

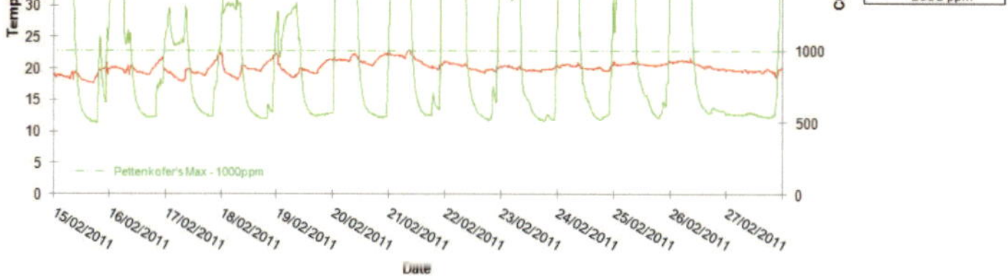

Figure 4. Unhealthy Bedroom Performance, SP1, House A, Bedroom 2

Another bedroom performing particularly poorly was House A, Bedroom 2, again during SP1 (Figure 4).Although levels of CO_2 were less frequently above 1000pm, compared with the previous example, it did have the highest level of CO_2 recorded over the whole investigation. The room was also occupied by one person but, it is worth noting that this bedroom was subject to approximately 30 hours of natural ventilation over the duration of the study period. The maximum level recorded was 2007ppm and the minimum level was 497ppm, giving an excessive range, of 1510 (Table 7). Air quality in relation to CO_2 levels was poor. Ta and RH data recorded max. and min. values of 22.9°C and 17.6°C, 59.3% and 38.1% respectively.

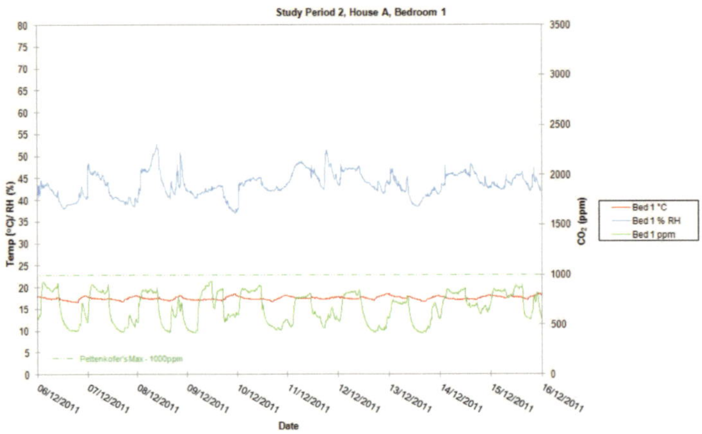

Figure 5. Healthy Bedroom Performance, SP2, House A, Bedroom 1

Figure 6. SP1, House A, Bedroom 1

The healthiest levels of CO_2 were recorded during SP2 in House A, Bedroom 1. (Figure 5) This room had one occupant and was not subject to any natural ventilation. Data recorded

http://dx.doi.org/10.5772/51900

for this bedroom was continuously below 1000ppm. The maximum level recorded was 939ppm and the minimum was 386ppm (Table 7). This produced an overall range in CO_2 levels of 553ppm. Importantly, the room appears to respond better to occupation that in previous examples, with CO_2 levels remaining within the prescribed limits.

With reference to the two latter examples, (Figures 4 and 5) both bedrooms are in House A and therefore have the same construction. They also have identical floor area and volumes, (Table 2) and appear to differ only in orientation. An influential disparity between the two examples is the study period in which the data was collected.

SP1	**Min (ppm)**	**Max (ppm)**	**No. > 1000ppm**	**Time > 1000ppm (hr& min)**	**Mean Time/Day > 1000ppm (hr& min)**
House A – Bed 1	713	1398	1847	153hr 55	11hr 50
House A – Bed 2	497	2007	1513	126hr 5	9hr 42
House A – Bed 3*	681	1316	197	16hr 25	1hr 16
House A – Bed 4	478	1176	711	59hr 15	4hr 33
Bedroom Total	2369	5897	4268	355hr 40	6hr 50
House B – Bed 1	585	1569	1317	109hr 45	8hr 27
House B – Bed 2	445	2006	690	57hr 30	4hr 25
House B – Bed 3	713	1819	2085	173hr 45	13hr 22
House B – Bed 4	550	1907	1220	101hr 40	7hr 49
Bedroom Total	2293	7301	5312	442hr 40	8hr 31
Total	44662	13198	9580	798hr 20	7hr 41
SP2	**Min (ppm)**	**Max (ppm)**	No. >1000ppm	Time >1000ppm (hr& min)	**Mean Time/ Day>1000ppm (hr& min)**
House A – Bed 1	386	939	0	0min	0min
House A – Bed 2	393	1151	467	38hr 55	3hr 54
House A – Bed 3	374	1044	352	29hr 20	2hr 56
House A – Bed 4	367	1300	265	22hr 5	2hr 13
Bedroom Total	1520	4434	1084	90hr 20	2hr 16
House B – Bed 1	417	1226	334	27hr 50	2hr 47
House B – Bed 2	460	1075	56	4hr 40	28min
House B – Bed 3	433	1478	774	64hr 30	6hr 27
House B – Bed 4	375	1055	18	1hr 30	9min
Bedroom Total	1685	4834	1182	98hr 30	2hr 28
Total	3205	9268	2266	188hr 50	2hr 22

Table 7. CO_2 Levels, SP1 and SP2 (*House A – Bed 3 was not slept in, see Table 4.)

With this in mind, an obvious difference can be shown between the two study periods by comparing the best performing bedroom from SP2 with its counterpart in the first study. (Figure 5 and 6). SP2 shows much lower levels of CO_2 than that recorded in the same bedroom in SP1. In fact SP1 results indicate unacceptable levels of CO_2 within all the bedrooms of House A and B (Table 7). In SP2 only two of the eight bedrooms recorded levels above 1000ppm on a regular basis (>3 hours a day). Further to this, SP2's maximum levels are not as high, and the minimum levels were also lower.

Although a crude representation, looking at the mean time per day that bedrooms spend above 1000ppm clearly strengths the visible difference shown in the graphs, between the CO2 levels in both study period.

4.3. Discussion

The most important inconsistencies are highlighted below and will be considered when discussing the results.

House A and B had different occupancy distributions

Natural ventilation occurred during SP1

Thermostat increase occurred in House B during SP2

MVHR system was serviced in the interlude following SP1

4.3.1. Air temperature

The temperature thermostats were possibly set too high in SP1. It is possible that the Ta recorded for SP1 House A would have been considerably higher if the occupants had not opened the windows in the bedrooms so frequently. The need to naturally ventilate to different degrees between House A and House B, but with similar temperature outcomes, supports the observation of a difference in IAQ between the two houses. Likewise, it can be assumed that the thermostats in SP2 were set too low, resulting in one of the occupants altering the settings in House B to compensate. This evidence further identifies a difference between House A and House B. Whether the difference in Ta is due to construction type, a variation in MVHR system or a study limitation, the unacceptable Ta experience highlights the need for a greater degree of occupant control in order to achieve the ideal sleeping temperature.

4.3.2. Relative humidity

RH levels remain constantly poor throughout both study periods and suggest that the MVHR is producing air that is too dry, 30-40%. The reason for this unacceptable RH is unclear.

4.3.3. Large diurnal range

A diurnal CO_2 range is normal and can be expected to be evident in the results due to the occupancy pattern of a bedroom, as described previously. It is however the size of the range

http://dx.doi.org/10.5772/51900

produced, such as that seen prevalent in SP1 (Figure 3) is a concern. The bedrooms have acceptable CO_2 levels when empty but when they are occupied they fail to adapt. This observation leads to an assumption that there is very little or no ventilation taking place during SP1. This conclusion is supported by unstable Ta and RH levels also presented in the CP. Reasons for poor ventilation could be attributed to the MVHR system not functioning as required to produce the sufficient amount of air changes per hour needed for each room. There was no CP prior to SP2 to gain clarification from, however, a reduction in the diurnal range in SP2 shows that better ventilation must be taking place.

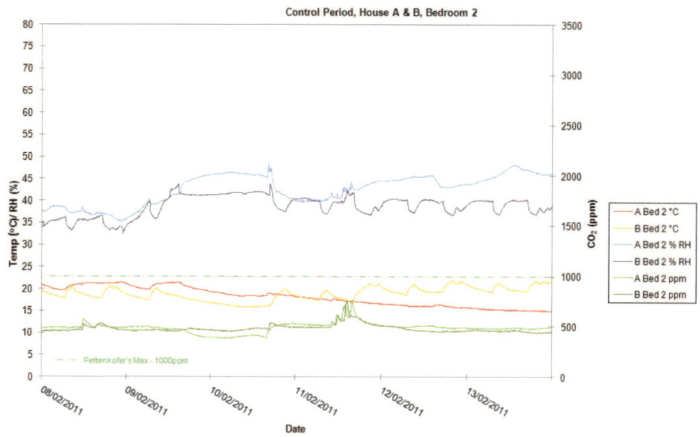

Figure 7. CP, House A & B, Bedroom 2

4.3.4. *Bedroom performance variation*

Other than the probable identification of a poorly performing MVHR system, there are other factors for varying bedroom performance.

As highlighted in Figure 3, Bedroom 3 in House B SP1 produces unhealthy CO_2 results. The room's size could be a contributing factor resulting in poor IAQ. Bedroom 3 has a volume less than half that of the largest bedroom (Table 2). It has no additional system design requirements, such as a larger supply vent or increased flow rate to provide more air changes. With the same occupancy levels the air will become polluted more rapidly than in a bedroom of a larger size, due to the smaller volume of air available for respiration.

It is also worth considering that the occupant slept with the door shut, magnifying the IAQ problem. There are no trickle vents between rooms and with the door shut there is no way of stale air leaving the room, this in turn inhibits supply flow.Bedroom 3's performance is of specific interest because it has the most potential to be used as a nursery. As previously mentioned in the IAQ of these spaces must be excellent as it is a time when childhood illness/IAQ related health issues are most common.

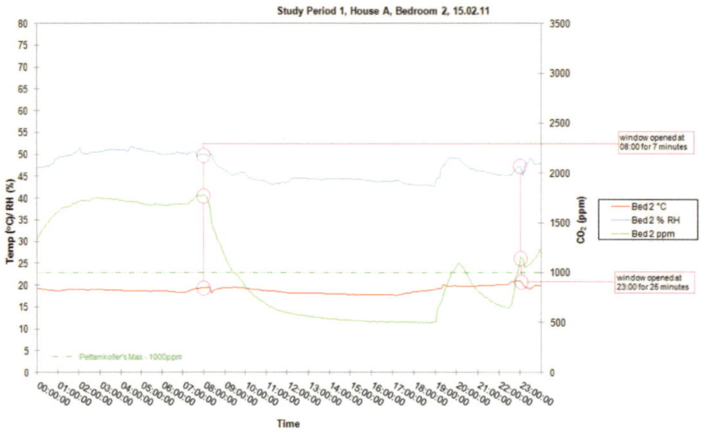

Figure 8. Window Opening, SP1, House A, Bedroom 2

In the same study Bedroom 2 in House A also performs poorly in relation to IAQ. Bedroom 2 has a volume of approximately 28m³ compared with 17m³ in Bedroom 3. This room was heavily ventilated naturally over SP1 (Table 3). By looking at how conditions changed during each recorded window opening period it is assumed that without this additional natural ventilation Ta and CO_2levels would have been considerably higher. Its counterpart in House B was only subject to roughly four hours of natural ventilation compared with 30, but performed considerably better, although maximum and minimum values were similar. It is possible that in House B the occupant slept with their bedroom door ajar opposed to that of House A. This is unconfirmed. The results produced for this room during SP1 would lean towards House B having producing more acceptable IAQ and are backed up by qualitative assessment which found occupants rated House B's IAQ to be superior to that of House A.

In addition to the examples discussed it is worth noting that the attic rooms seem to perform relatively well in each case, even with House A having double the occupancy level. This could be due to Bedroom 4's large volume and also its proximity to the MVHR unit resulting in shorter duct lengths and therefore better flow rate. The room's location near to the unit may also explain why in the qualitative assessment made by an occupant it was stated that the fan was very obvious and noisy in the attic bedroom. With this is mind, the position of MV supply vents and the noise levels they produce should be considered to the same extent, especially as the air tight construction results in the home being extremely quiet in general.

4.4.4. Study period variation

The substantial difference shown in the results, between the chosen examples, highlights a clear improvement in CO_2levels between SP1 and SP2.The qualitative IAQ results recorded support the quantitative results. The inhabitants deemed the air to be of greater satisfaction, freshness and circulating more frequently within the second study period.

http://dx.doi.org/10.5772/51900

Significant evidence explaining reasons for the improvement in CO_2levels can be sought from a report written following an inspection of the MVHR system in both houses, by the installer, subsequent to the completion of the first study period. The report identified a number of defects, which were rectified between SP1 and SP2.

In House A it was observed that 125mm ducting had been used in the roof space, but that the majority of ducting installed was only 100mm. This potentially led to the unit running at higher pressure than it was designed to, although no increase was specifically noted. In House B, however, the system was measured to be running at an increased pressure. Higher pressures impact on air flow and created the potential for fans to stall, resulting in reduced ventilation.

There were several reasons for the high pressure within the system of House B. As well as the extensive use of 100mm ducting, there were also additional bends compared to that of the ducts in House A. In addition the ducting had been connected in the wrong position to the MVHR unit. This contributed to the high pressure as well as reducing the performance of the system. Areas of ducting were also found to contain remnants from the internal fix out of the house, hindering air flow. Other ducts had been squashed. Both faults increased pressure within the system. The filters within the unit were seen to be dirty and in need of cleaning/replacement. This factor would have resulted in increased resistance adding to the high pressure within the system. The build up of dirt would have reduced the filters air purifying efficiency resulting in increased levels of contaminants being circulated throughout the house. Extract flow rates in the rooms were measured at low levels because the system was running poorly, this would result in polluted air being removed from rooms at a much slower rate than necessary during inhabitation.

It seems reasonable to conclude that these works have improved the performance of the system, but that concerns remain in regard to the levels of IAQ that are experienced by occupants, particularly during peak conditions.

5. Case study 2: Gilmores close

This project was a building performance evaluation of an adaptive rehabilitation project on a Category B listed 19th Century stone tenement located within the World Heritage Site of Edinburgh's Grassmarket. Working within the constraints of its historical significance and limited budget (a registered social landlord as Client) and end user group, this project has sought to create an energy efficient solution for its sustainable rehabilitation.

To assess the performance of this building MEARU undertook a programme of monitoring and evaluation over a three-week period during March 2011(from 17.03.11 to 12.04.11). Average external temperatures during this period were 9 °C. Environmental monitoring was supplemented with an analysis of energy demand and acquisition of qualitative data through semi-structured interviews of the occupants, and observations by the surveyors to provide an overview of building performance.

This project was undertaken over a limited, albeit very focused, period. As such the information derived provides a 'snapshot' of building performance, rather than a more extensive review of performance over the course of an annual climatic cycle. The study collected data on 6 properties (5 dwellings and 1 small office) out of a potential 17 properties.

5.1. Construction

The measures used in the refurbishment of the block incorporate specific approaches to design and specification to reduce the on-going environmental impact of the building and to improve the living conditions of the potential residents. Working within the constraints (both physical and statutory) of the existing blonde ashlar and random rubble sandstone façade and structural cores, a new internal layout was constructed to provide flatted accommodation. The new insertions within this masonry skin are generally lightweight timber construction. Figure 9 shows the general flat arrangement.

Figure 9. Typical Plan Gilmore Close

The thermal performance of the building was improved by bringing the fabric up to contemporary standards through a process of internal dry lining and insulation to achieve a U-value of 0.25W/m^{2}K.The thermal performance of the historic timber sash and case windows was also improved through the use of secondary internal glazing improving U-values to 1.8W/m^{2}K. Both of the above strategies adhered to the design principles dictated by the building's historic status in that they did not materially affect the principal elevation. To the rear a south facing, semi-glazed (approx. 50%) sunspace with an average U-value of circa 1.0W/m^{2}K has also been incorporated into 12 of the dwellings to provide additional amenity and to make use of passive solar gains.

The principle active technology employed throughout the development is a vertical ground source heat pump (GSHP), which, along with an electric back up heater, provides for the hot water and space heating demands of the full building. Delivery of the space heating is through a wet under-floor heating system. Due to limitations of the timber intermediate floor structure this is provided within proprietary insulated trays rather then being contained in a screed.

Ventilation of 13 of the dwellings also allows for the use of heat recovery through proprietary mechanical ventilation with heat recovery (MVHR) units. In the 1 bedroom apartments (without sun spaces) a conventional system of opening windows, background trickle ventilators and mechanical extraction from wet spaces has been installed. Elsewhere a whole house MVHR system draws air from the kitchens, bathrooms and sunspaces, and after recovering waste heat, delivers fresh warmed air to the hall spaces, with the intention that this will dissipate to adjacent spaces. Note that MVHR relates to an energy strategy but is viewed primarily as a ventilation aspect with the heat recovery aspect being secondary.

5.2. Occupancy

The building has three distinct groups of occupants, all of who were represented in the data collection process. The first user group is that of the mainstream social rent tenants. They occupy one of the building's two closes. The second user group, occupying the second close, is made up of residents who require supported living. The third group of users are the care staff who occupy the building's office space and provide support to user group two.

Following completion and occupation of the building there were reports from residents of poor performance and problems with the heating system. Through a process of further commissioning and alteration this system was brought up to a standard where resident complaints were dramatically reduced but where continued problems were evident. Anecdotal evidence suggested over-heating was common and this was be supported by visual inspections of window openings.

In response to these issues MEARU was asked by the architects to undertake a building performance evaluation to identify issues relative to the building performance in general with a specific focus on internal comfort. The project was funded by the CIC Start Online academic consultancy fund. The key question posed was what energy performance and environmental conditions are being achieved, and if these are below requirements what lessons may be learned for this, and other similar projects.

5.3. Data review

Research into the building performance and user satisfaction was undertaken using a variety of approaches and techniques for data collation and analysis. This was designed to primarily provide a resource of quantitative (empirical) data but which was supported by qualitative data providing a greater depth to the analytical process. Over a 3½ week period the internal temperature, relative humidity and CO_2concentration were monitored in all apartments, the hall and kitchens of five flatted dwellings and throughout one office space (noting that in each case the bathrooms/ WCs were omitted).

Although not a longitudinal study, there are significant benefits in a short, intense period of monitoring. The relatively brief duration led to limited intrusion on the occupants, ensured continuity in data collection relative to both dwellings and occupants and allowed a fine granularity, which helped to identify specific events within the flats.

5.3.1. *Thermal comfort*

Due to the anecdotal evidence on overheating, this was the initial focus of initial research. A review of physical data at the macro level (Table 8) confirmed that the mean and absolute maximum temperatures within all apartments (office space excluded) were - often significantly - beyond the accepted comfort range. The mean values confirmed the suspicions held at the project outset but did not provide any information on cause or potential solutions.

Room	Mean Temp (°C)	Comfort Temp (°C)	Δ T 1 (°C)	Absolute Max (°C)	Δ T 2 (°C)
Living Rm	22.62	21.00	+1.62	28.00	+7.00
Kitchen	22.87	18.00	+1.87	29.10	+11.10
Hall	23.45	18.00	+5.45	31.20	+13.20
Sun Space	21.24			40.90	
Bedroom 1	22.58	18.00	+4.58	27.20	+9.20
Bedroom 2	21.41	18.00	+3.41	26.20	+8.20

Table 8. Mean and absolute maximum thermal conditions over project duration

To identify this, a more focussed review was undertaken of each dwelling relative to the profile of physical parameters on a diurnal basis.

Figure 9 illustrates a typical daily example where a living space is heated to a degree of discomfort and then is rapidly cooled by the occupant behaviour of liberal window opening. This behaviour was found to be repeated throughout the development and was supported by the survey responses in which 60% of residents noted they opened windows every day throughout the year.

Recorded data from an unoccupied dwelling had shown that a relatively stable temperature profile could be maintained internally which demonstrated that despite the loss of thermal mass necessitated by insulated dry-lining and timber construction the fabric was capable of facilitating thermal comfort. Further investigation using thermal imaging provided an insight to problems of frequent overheating. Figure 11 shows the surface temperature of a typical apartment floor at two different points in time. In the first (T1) the thermostat was set at its lowest level yet a temperature of 28.9°C was evident. Immediately after this image was taken the thermostat was turned to it's highest setting with the same image being taken one hour later (T2).

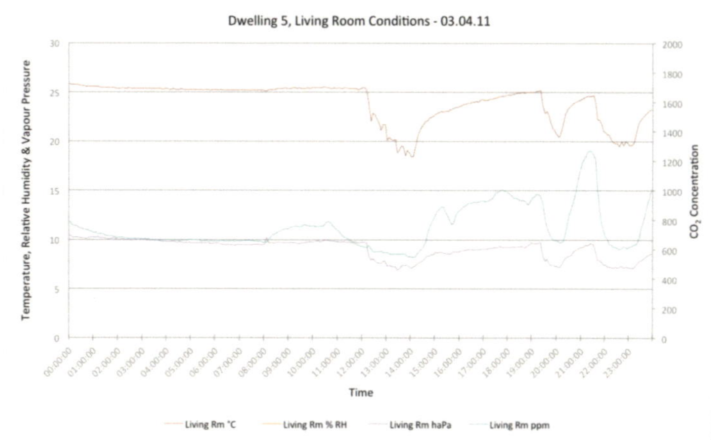

Figure 10. Physical parameters in Dwelling 5 living room – fluctuating thermal comfort

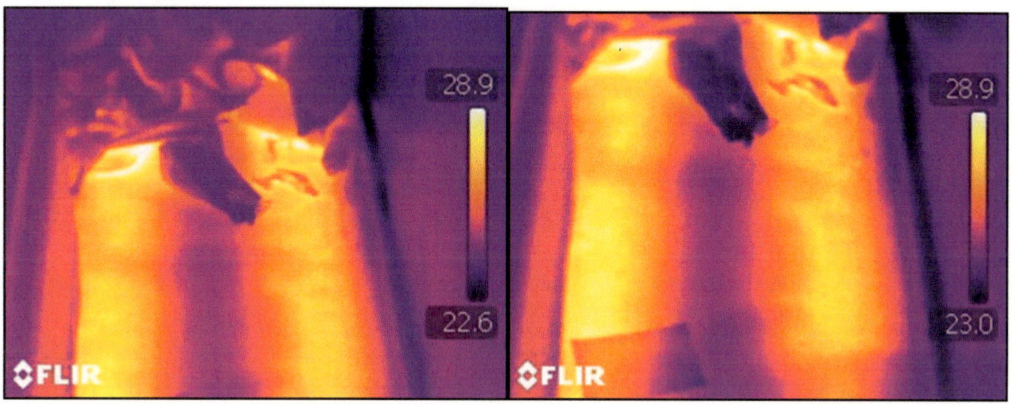

Figure 11. Floor surface temperature T1 and T2

The level of the initial reading suggested that the control of the heating system was ineffective. This was confirmed by the lack of response over the subsequent sixty-minute period. Poor performance of heating controls, allied to a poor user interface, were identified as factors that consistently resulted in the creation of sustained internal temperatures exceeding the comfort range. In addition to this, the lack of thermal mass in the structure, an outcome of the approach to thermal upgrade of the historic fabric, results in high rates of heat gain and loss; a process which is difficult for residents to stabilise once the cycle of window opening has commenced. Ultimately, having windows open when heating is on leads to an increase in the energy required for space heating and undermines the thermal efficiency of the building. This also provides an explanation for the disparity in predicted and measured energy loads for space and water heating.

5.3.2. Internal air quality

Monitoring identified several spaces with very good IAQ. Given the prevalence of window opening this result was hardly surprising, but will of course have a thermal penalty. In circumstances where window opening is common, the use of an MVHR system is not only ineffective but is also an additional primary energy burden on the dwelling as the fan continues to run at the same rate regardless of IAQ conditions. Where window opening was not prevalent, maximum values of CO_2concentration were frequently found to rise and be sustained above recognised maximum desirable levels of 1000ppm. Figure 11 illustrates a particular situation from a bedroom in Dwelling 2 over the monitoring period but this is typical in monitored apartments throughout the project.

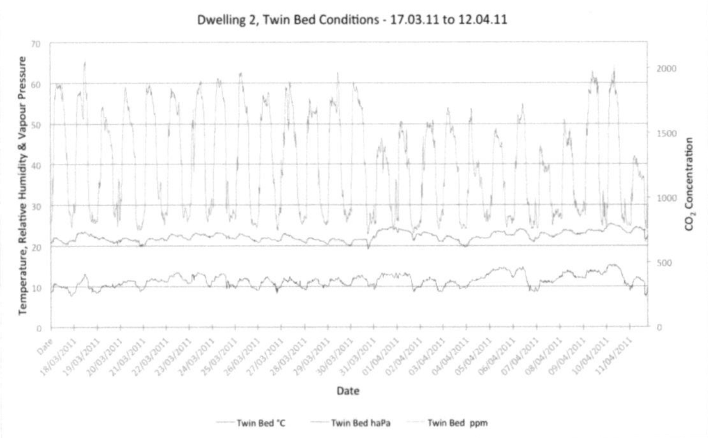

Figure 12. Physical parameters in Dwelling 2 twin bedroom – high CO2 concentrations

The bedrooms are of particular interest as windows are generally closed at night due to issues of external noise and as the occupants are asleep no further occupant intervention occurs. Accordingly these spaces are entirely reliant to the MVHR system to deliver ventilation and good IAQ.

5.3.3. Discussion

A contributory factor to the poor performance is the layout of the system, with the two air delivery registers focussed into the hall space only.The intention behind this design was that air would migrate from this central location into adjacent apartments. However, fire regulation requirements for self-closing fire doors and smoke seals prevent such air movement. The situation appears to have been exacerbated post-construction, as carpet installation has further reduced the air spaces under doors.

With the potential health impacts, the importance of good IAQ cannot be a secondary concern and it must not be undermined by attempts to improve thermal efficiency and air tightness. Not withstanding this position, the level at which poor air quality is perceptible has the potential to cause occupants to manually seek improved ventilation. With a CO_2concentration of 1000ppm poor air quality is perceptible to humans with the stress initiated behavioural response invariably being one of window opening and the result being, as was evidenced with the poor thermal control, one of high energy loss. Instances of this were identified through in the monitoring of this project and the outcome of poor air quality is (entirely rational) behaviour that counteracts the approach to energy conservation central to a contemporary design ethos.

A further issue of note in relation to the MVHR system is that of maintenance. The MVHR system contains air filters to screen for dust and particulate and the need ensure these are cleaned on a regular basis is critical to the functionality and energy consumption of the system. In this instance limitation on the space available has led to the placement of the unit in a location above a dropped ceiling where access for maintenance and filter replacement is physically very difficult.

The study identified that through the refurbishment considerable improvement has been made in the thermal performance of the buildings and there should be no doubt that overall it has been successful in terms of the improvement of the flats and the maintenance of the cultural heritage of these buildings.

However, the study also found that there are some problems that lead to a reduced energy performance and some unintended negative consequences, particularly in respect of indoor air quality. These are due to some design issues, for example the fire protection measures over-riding the ventilation strategy; the desire to increase living space in the sun-spaces undermining the energy strategy (not discussed in this Chapter); the loss of thermal mass through the provision of internal insulation and lightweight sub-division; and the design integration of elements such as the MVHR and sunspaces. There are also issues of installation, commissioning and maintenance, including the lack of proper control of the heating system, and MVHR specification and installation.

The study identified a number of measures for improvement, both remedial measures in these flats, but also lessons for similar developments elsewhere.

Options for improvements in this development that are currently being explored include: the re-commissioning and improvement of the heating control systems; an extension of the MVHR supply ducts to deliver air directly into living spaces and bedrooms; provision of additional control over the sunspace extract system.

These findings are also relevant to future developments in this building type. The original building would have had an energy strategy relying on open fires with chimneys providing significant levels of radiant and convected heat, which would engage with the thermal mass of the building. Sash and case windows, although relatively draughty by contemporary standards, provide high and low level openings, which, when combined with high ceilings give very good ventilation regimes. The literature review for this project has highlighted a significant gap in the understanding of the standards of IAQ in energy efficient dwellings and this is a key area for further study. This is relevant to new build energy efficient dwellings and particularly to retrofit schemes as contemporary approaches may actually reduce IAQ and undermine attempts to improve thermal efficiency and reduced CO_2output.

The necessity of removal of some of these characteristics (thermal mass, volume of dwellings through sub-division, high ventilation rates) needs to be considered in taking a holistic approach to refurbishment of these dwelling types which addresses environmental performance as well as energy targets.

These questions have wider implications for the profession and identify areas for further research if we are to achieve the desired sustainable future.

6. Conclusions

Both these projects clearly illustrate the challenges for sustainable energy futures that can arise if environmental strategies are not successful. Both of these projects have produced low energy buildings, of high architectural quality, but these studies have identified that unless environmental strategies are carefully designed, implemented and maintained, unintended negative consequences can arise. As well as producing potentiall harmful environments which can have detrimental effects of occupants health, poor ventilation design can undermine strategies for energy conservation as occupants attempt to achieve comfort by conventional means such as window opening.

In both projects the role of building users is two-fold – as operators and as consumers. In the former role, building users have a part to play in ensuring the efficient and effective performance of the buildings, but this role is significantly moderated by their environmental experience, the effectiveness of the systems and their control over them. Thus a system that cannot effectively control comfort leads to behaviours that can increase energy consumption. In the latter role, building users are subject to the environmental conditions of the buildings and may experience discomfort or even detrimental health effects if the environmental strategies are ineffective.

The results and discussions from both projects conclude that the design of mechanical ventilation heat recovery systems does require further consideration in order to provide healthy levels of IAQ in both projects whilst maintaining good energy efficiency. In the Glasgow House the considerable improvement in CO_2levels between the study periods indicates that the design itself is reasonably good, however, there is still need for improvements within the bedroom areas and poor IAQ was still experienced. Remedial measures are also proposed in Gilmores Close and it is hoped that there will be an opportunity to monitor the results of this. The projects also illustrate the need for holistic environmental design, which takes into account a range of design requirements, including obvious environmental factors such as heating, ventilation, lighting, and controls, but also relating these to other design parameters such as the nature of occupants, maintenance and fire safety.

The studies also illustrate the value of information revealed by the use of BPE. Underperforming systems which potentially endangers the health of the occupants will not be apparent unless the buildings are evaluated in use. Without these studies the conditions in these dwellings would not be known, nor would the knowledge about the issues and how they may be addressed be revealed to the industry in general. In the current low carbon environment most new buildings are experiments – it is vital that we go back and check the results. Collectively, the design, installation, maintenance and the need for BPE require further consideration in order to provide exemplary indoor air quality in Scotland's future housing stock.

Author details

Tim Sharpe

Mackintosh Environmental Architecture Research Unit, Glasgow School of Art, Glasgow, UK

References

[1] Sullivan, L. (2007). A Low Carbon Building Standards Strategy for Scotland, Arcmedia, Scotland.

[2] Shorrock, L. D., & Utley, J. I. (2003). Domestic Energy Fact File 2003. BRE Bookshop, Watford, England

[3] Boardman, B. (2007a). Home truths: a low-carbon strategy to reduce UK housing emissions by 80 per cent by 2050. Report to the Friends of the Earth and the Cooperative Bank.

[4] Steemers and , K., & Yun, G. Y. (2009). Household energy consumption: a study of the role of occupants', Building Research and Information, doi.org/10.1016/j.enbuild.2010.09.021., 38, 625-637.

[5] The Rebound Effect: an assessment of the evidence for economy-wide energy savings from improved energy efficiency.A report produced by the Sussex Energy Group for the Technology and Policy Assessment function of the UK Energy Research Centre, Steve Sorrell, October (2007). 1-90314-403-5

[6] http://www.cicstart.org/content/home/1,1,276/FS01DevelopmentofPostOccupancyEvaluationforevaluationofinnovativelowcarbonsocialhousingprojects.html

[7] Energy, , Buildings, Volume. 4., Issue, ., & January, . (2011). Measured energy and water performance of an aspiring low energy/carbon affordable housing site in the UK, 117-125.

[8] Mazzeo, Nicolás A, ed., Chemistry, Emission Control, Radioactive Pollution and Indoor Air Quality, (Croatia, 2011), p. 447

[9] M. Jantunen, J. J. K. Jaakkola and M. Krzyzanowski, eds., "Assessment of Exposure to Indoor Air Pollutants," in WHO Regional Publications, European Series, no. 78, (1997), p. 1

[10] Raymer, Residential Ventilation Handbook, p. 2.4, McGraw Hill, ISBN: 9780071621281
2009

[11] Mazzeo, ed., Chemistry, Emission Control

[12] Stirling Howieson, Housing and Asthma,Oxon, (2005).

[13] Jantunen, Jaakkola and Krzyzanowski, eds., "Assesment of Exposure to Indoor Air Pollutants,"

[14] Allard, ed., Natural Ventilation in Buildings

[15] Scottish Government, Scottish Domestic Building Regulations, Section 3.14

[16] Porteous, "Sensing a Historic Low-CO2 Future," p. 238

[17] Allard, ed., Natural Ventilation in Buildings

[18] Michael G. Apte, Joan M. Daisey, William J. Fisk, "Indoor Carbon Dioxide Concentration and SBS in Office Workers," in Proceedings of Healthy Buildings(2000). http://senseair.hemsida.eu/wp-content/uploads/2011/05/6.pdf, 1

[19] Dearden, "Ventilation and Air Quality," p. 21

[20] Mosley, "Fresh Air and Foul," p. 8

[21] Ibid., p. 21; London Hazards Centre Handbook, Sick Building Syndrome, p.43

[22] Crump, D., Dengel, A., & Swainson, M. (2009). Indoor Air Quality in Highly Energy Efficient Homes- a Review, IHS BRE Press, Watford, England.

[23] Platts-Mills, T. A. E., & De Weck, A. L. (1989). Dust Mite Allergens and Asthma- A worldwide Problem, Journal of Allergy and Clinical Immunology, , 83, 416-427.

[24] Heinrich, J., Int, J., Hyg, Environ., & Health, . (2011). Jan;Epub 2010 Sep 18.Influence of indoor factors in dwellings on the development of childhood asthma., 214(1), 1-25.

[25] Arundel, A. V., Sterling, E. M., Biggin, J. H., & Sterling, T. D. (1986). Indirect Health Effects of Relative Humidity in Indoor Environments,. Environmental Health Perspectives, , 65, 351-361.